Table of Contents

Message from the Family Liaison Office

Dear Evacuee:

Welcome back. Your last few days have probably been difficult and you have not had a chance to plan or think things through. This evacuation information was put together by the Family Liaison Office as a resource and guide to help you settle into life in your safehaven.

The Family Liaison Office (FLO): FLO is here to assist as you try to bring some order back into your lives. We provide services in three areas:

> *Attention evacuees:* **To receive information pertinent to your post's evacuation status, please provide the Family Liaison Office with your contact information, including email address and phone number as soon as possible. The Family Liaison Office is responsible for maintaining contact with evacuees.**

- *The Crisis Management and Support Officer* and *Program Assistant* are your principal FLO points of contact. They can help you navigate the world of allowances and regulations. They will also remain in contact with your community during the entire evacuation.

- *The Employment Program Coordinator* can provide information for family members on employment and other opportunities in the DC area.

- *The Education and Youth Officer* can assist with your children's educational needs in the DC area and with information on boarding schools throughout the country.

Please feel free to call or stop by to meet with the FLO staff. Note that as of mid-January, 2006, unless you have a valid Department of State Smart ID, you must **enter the building through the C Street (diplomatic) entrance**. All visitors must be screened in exterior security screening facilities at the C Street lobby for clearance into the building. Our contact information follows:

> Harry S Truman Building (Main State)
> 22nd and C Street NW, Room 1239
> Washington, DC 20520
> Telephone: (202) 647-1076, or toll-free (800) 440-0397
> Fax: (202) 647-1670
> Email: flo@state.gov
> For Evacuation Questions: FLOAskEvacuations@state.gov
> Website: http://www.state.gov/m/dghr/flo

The Employee Consultation Service (ECS): is a **confidential** counseling service available to evacuees and their families. The office is staffed by clinical social workers who meet with clients by appointment, on a drop-in basis, through e-mail or by phone. Many evacuees have found it helpful to talk with them about difficulties their children are experiencing handling a sudden move as well as about their own reactions. The clinical social workers are available for individuals and/or group counseling.

> Employee Consultation Service
> Columbia Plaza, SA-1, H-246
> Telephone: (202) 663-1815. FAX: (202) 663-1456
> Email: MEDECS@state.gov

The Transitions Center at the George P. Schultz National Foreign Affairs Training (FSI): offers many courses of interest to Foreign Service family members. To register, email FSITCTraining@state.gov with your name, address, agency, and social security number. Indicate that you are an evacuee.

> The Transition Center,
> George P. Schultz National Foreign Affairs Training Center
> 4000 Arlington Blvd., Arlington, VA 22204
> Telephone: (703) 302-7268
> Fax: (703) 302-7452
> Email: FSITCTraining@state.gov
> Website: http://www.state.gov/m/dghr/flo

Please call the Family Liaison Office as soon as possible (at (202) 647-1076 or 1 (800) 440-0397 or email flo@state.gov) to let us know your safehaven address, telephone number(s), and your email address. We would like to stay in contact with you throughout your time in the United States and need to be able to contact you quickly if there is a change in the evacuation status. If you are in the Washington area, we will also arrange occasional briefings to provide opportunities to hear what is happening and to ask questions on both political and administrative matters. We can also help you process the financial paperwork for your evacuation. Once again, welcome back. We look forward to assisting you in any way we can.

Sincerely,

Leslie Teixeira, Director

Evacuation Definitions

The primary purpose of the evacuation of a U.S. Mission is for the safety and security of Mission personnel and their families. The following definitions are in accordance with Department of State regulations. Please note that the military uses slightly different definitions for some of these terms.

Authorized Departure: This type of evacuation is voluntary for official family members and non-emergency direct-hire employees. It allows the Chief of Mission greater flexibility in determining which employees or groups of employees may depart. There is no difference in benefits between authorized and ordered departure. *The employees and family members who choose to depart post on authorized departure may not choose when to return to post.* They may not return until the Under Secretary for Management has lifted the evacuation.

Ordered Departure: An Ordered Departure evacuation is not voluntary. Family members and non-emergency staff are ordered to depart post on evacuation status. It is not uncommon for the status of an evacuation to shift from authorized to ordered, depending upon the situation at post. In some evacuations, all staff must leave and operations are temporarily suspended at post. Members of Household (MOH) are entitled to evacuation assistance, but cannot be ordered to depart. However, it is important to note that if an MOH is occupying USG housing, and the Chief of Mission (COM) feels that the presence of MOHs in post housing could impact post security resources or otherwise affect post operations, then the COM can direct the employee to move MOHs out of USG housing. In addition, if the MOH remains at post, they would not have access to official post facilities and services while the employee remains on evacuation.

Drawdown: Refers to the evacuation of family members and non-emergency staff, leaving emergency personnel at post. One of the tasks of the Emergency Action Committee at post is to determine which staff positions would be necessary during a crisis. In an evacuation, the post would be **drawn down** to that number, unless the crisis required the temporary closure of the post.

Safehaven: When personnel and family members are evacuated from post, the official safehaven destination is the United States. Employees must report to their agency headquarters, while family members may choose a safehaven location anywhere in the United States and Puerto Rico; they are not required to return to their home leave address. Employees may request an alternate foreign safehaven for family members, however, the request must be approved by the Under Secretary for Management.

Subsistence Expense Allowance (SEA): The allowance given to official evacuees, based on locality per diem.

Meals & Incidental Expenses (M&IE): Part of the Subsistence Expense Allowance, M&IE is a flat rate not requiring receipts (amount varies depending on locality).

Commercial lodging: Any temporary lodging, such as a hotel, for which one pays a rent and can produce a receipt.

Non-commercial lodging: A lodging for which one does not pay a receipted rent, such as one's own home or a relative's residence.

Separate Maintenance Allowance (SMA): An allowance available to employees whose official family members have not chosen to accompany the employee for a tour of duty at a particular post. The employee may be eligible for *Involuntary Separate Maintenance Allowance (ISMA)* when, after an evacuation, family members are not allowed to return to post but the employee does.

Transitional Separate Maintenance Allowance (TSMA): Following the termination of an evacuation and the conversion of a post to an unaccompanied status, or for educational reasons following the termination of an evacuation and reversion of post to accompanied status, the employee may be eligible to receive TSMA. Consult the Department of State Standardized Regulations 262.3 for a complete explanation of the eligibility requirements of these benefits.

Choosing Your Safehaven: Factors to Consider

In an evacuation situation, family members may choose a safehaven at any point in the United States, including Alaska, Hawaii, and Puerto Rico. They do not necessarily have to choose their home leave address. When deciding on a safehaven location, family members may want to consider the following options, and the pros and cons of each.

THE METROPOLITAN WASHINGTON, DC AREA

If the employee is evacuated, she or he **must** report for duty at the agency headquarters, usually Washington, DC. In this case, family members may wish to choose Washington as their safehaven point. Family members earlier evacuated to a U.S. or authorized foreign safehaven may be permitted to rejoin an employee subsequently evacuated and reporting to duty in Washington, DC. Being in Washington has certain advantages, including being able to come into the State Department for briefings organized by the Family Liaison Office (FLO) or the evacuated post's Community Liaison Office Coordinator working out of FLO; taking courses at the Overseas Briefing Center; and having the resources of the Department near at hand. If a group of evacuees is located in the Washington area, they have the opportunity to enjoy mutual support, and an evacuee may feel a little more in touch with the situation at post.

OTHER LOCATIONS WITHIN THE UNITED STATES

Often the most important factor for evacuees in making the safehaven decision is where their network of support is located. If family and friends are located on the other side of the country from Washington that may well be the best safehaven for the evacuee. If the employee did not return to the U.S. on evacuation status, it may be even more important to be near a source of family support, particularly if there are small children involved. Although FLO makes every effort to keep in contact with all evacuees wherever they are located by phone or e-mail, those located farther from Washington have less access to certain resources. However, this is often less important to evacuees than the support of their family, friends and hometown.

ALTERNATE FOREIGN SAFEHAVEN

Although the official safehaven for evacuated families is the U.S. and Puerto Rico (in rare cases the official safehaven may be in a foreign country), some family members may not have particular ties to the U.S. and would prefer to choose a foreign location. **Note that the post must request an alternate foreign safehaven for a family member from the Department of State**. If the request is approved, the evacuee will receive a Subsistence Expense Allowance based on the lowest of the following: the official safehaven, approved alternate foreign safehaven, or the standard CONUS rate. If the request was not approved, the evacuee will not receive travel, evacuation or education benefits.

CHANGING LOCATIONS

Once the evacuee has arrived at the safehaven location, he or she will not be funded for travel to another location during the period of the evacuation. One exception: Funding for travel is allowed if the travel is to relocate and rejoin an employee newly returned to the United States. Family members who move from one location to another during the evacuation may do so at personal expense. SEA payments will then be based on the new official safehaven.

Coping Strategies for Evacuees

Evacuations elicit a variety of different feelings, but the universal response to an evacuation is a sense of not being in control of one's own life. The individual feels powerless, caught in a situation which affects every aspect of life. Since this feeling is so common, the following tips are suggested as ways to gain a measure of control over the situation.

MAKE CONTINGENCY PLANS

Decide ahead of time on a safehaven location, organize the documents to take to post, make plans for the children, MOHs, pets, and have powers of attorney (including one accepted by your bank or credit union) in order. Keep and use a copy of FLO's *Contingency Plan: Don't Leave Home Without It*. This document can be found online at http://www.state.gov/m/dghr/flo

PLAN FOR THE LONG TERM

Evacuations average 3 - 4 months. While the length of any evacuation is difficult to predict, those who plan for a longer, rather than a shorter, period of time experience fewer frustrations.

USE RESOURCES

While in the Washington, DC area, take some courses at FSI's Transition Center http://www.state.gov/m/fsi/tc/, including the many long distance training courses available. Consult with the executive office of your bureau and the FLO employment staff http://www.state.gov/m/dghr/flo/c1959.htm about short-term employment. The licensed clinical social workers at the Department of State's Employee Consultation Service at (202) 662-1815 (or their equivalent in other agencies) may be helpful. Also, the Department of State, USAID and many other USG agencies have contracts for employee assistance programs through Information Quest, a free resource and referral service available around the clock. The phone number is (800) 222-0364 or (888)262-7848 for the hearing impaired). The website is: www.WorkLife4You.com.

CREATE A "NORMAL" LIFE

Develop as normal a routine as possible for yourself and your children. Try to eat healthy food, get adequate exercise and socialize with family and friends. If an evacuation lasts more than a month, you may choose to put the children in school; some overseas schools have virtual programs so the children can continue their education with the same institution.   Get them involved in activities, and get involved yourself. Pursue hobbies, do volunteer work, or take a part-time job.

KEEP IN TOUCH

Stay in touch with fellow evacuees, with FLO, your CLO, and/or your assigned point of contact (i.e. the family liaison specialist for your agency) throughout the evacuation. You'll be up-to-date on the latest information from post, and enjoy mutual support with others in the same situation. FLO (or your agency's family liaison representative) will phone and/or send e-mails regularly, providing information on the status of the evacuation, and other important news. With permission, FLO will share with fellow evacuees your phone number and email address so that you may stay in contact with friends from post. FLO encourages posts to permit the CLO Coordinators to work out of the FLO Office during an evacuation to help maintain a sense of community between the evacuees and the post. Together, FLO and CLOs may organize Town Hall Meetings in the Department of State for employees and family members from all affected agencies.

There are instances when evacuees may not return to post at all. This can be a very difficult time for many as they have left their homes, their friends, and the life they knew in that country without having said goodbye. There is a sense of "unfinished business" about their post. Many experience an emotional loss. Most people who experience an evacuation are able to put it into perspective and go on, yet Foreign Service life never seems quite the same again. The experience may make evacuees more wary, and hopefully, will influence them to take contingency planning more seriously in the future. Eventually, the memory of an evacuation becomes part of the rich tapestry of experiences, positive and negative, which make up the life of a Foreign Service family.

Pets in an Evacuation

The official policy of the Department of State is that it does not evacuate pets. We recommend that, as a part of evacuation contingency planning, one make several optional plans for the pet so that one is ready for any situation that might arise in an evacuation. These options might include:

- Think about sending the pet out of country ahead of time if there seems to be a likelihood of evacuation, and if there is time. Identify someone back home to whom you can send the pet.

- Identify someone in-country with whom you can leave the pet if there is a crisis and you cannot take the pet with you. Make arrangements with this person now so that you are prepared in the event of an evacuation.

- Prepare to send your pet out of the country using a commercial carrier. Make sure shots are up-to-date and a pet carrier is ready, should you need to ship your pet out quickly. Be informed about pet travel regulations of host country and any country you must transit. Be aware, again, that you are responsible for the cost of the pet's evacuation.

- FLO suggests that all these options be in place at once, since one cannot predict what form an evacuation will take. Keeping shots up to date, your pet in good health and an appropriate carrier ready will be important in all these scenarios. The Overseas Briefing Center has country-specific information on quarantines and import requirements. OBC also has lists of professional pet shipppers.

- Other Web sites of interest regarding pets include:
  http://www.clubpet.com
  http://www.interpetexplorer.com
  http://www.petswelcome.com
  http://www.travelpets.com

Evacuation Allowance Overview

A Summary of Department of State Regulations for Non-Military Employees and Eligible Family Members under Evacuation Orders

*The **Department of State Standardized Regulations** apply to all government civilians in foreign areas but each agency may have further implementing regulations. Check your agency's regulations before applying the DSSR (for Foreign Affairs agencies consult the FAM)*

PERTINENT REGULATIONS

Complete authorizing regulations are in Chapter 600, Standardized Regulations (Government Civilians, Foreign Areas), and may be supplemented by individual agency regulations and procedures. Additional Department of State guidance on evacuation benefits is provided in Chapter 1400 of the Department of State's Emergency Planning Handbook (12 FAH-1).

The authorizing official for evacuation benefits is determined in accordance with each agency's practices and procedures. In the Department of State, the managing geographic Bureau and the Bureau of Resource Management normally will determine this. Any official approving a benefit not clearly authorized by the regulations is responsible for seeking the advice of an appropriate authority.

PURPOSE

The purpose of evacuation benefits is to help offset added expenses incurred as a result of an evacuation/authorized departure. The employee continues to be responsible for normal family living expenses.

SAFEHAVEN

The United States, including Alaska and Hawaii and the territory of Puerto Rico, is the official safehaven for eligible family member evacuees, and Washington, DC or other U.S. duty station is the official safehaven for all employees. Employees will be notified of the officially designated safehaven location in any emergency requiring evacuation. While an alternate foreign safehaven location may be approved in individual cases by the Secretary of State (who delegates the authority to the Under Secretary for Management) when considered necessary and in the best interests of the government, evacuees are not entitled to the usual diplomatic privileges, immunities or administrative services. No evacuation benefits are available for evacuees who proceed to an unauthorized point. If a request for an alternate foreign safehaven is not approved, evacuation benefits will not be authorized until the evacuee arrives at the official safehaven.

TRAVEL EXPENSES

Employees are generally authorized travel only to Washington, DC or other U.S. duty stations.

Eligible family members (EFMs) may travel to the officially designated safehaven location at U.S. Government (USG) expense. If the U.S. is officially designated as the safehaven, travel will be paid to any point in the U.S. In the rare case an alternate safehaven is approved, travel will normally be reimbursed only on a cost-constructive basis calculated from the evacuated post to the nearest point of entry in the continental U.S. Any subsequent travel during the evacuation is at personal expense, except for authorized return travel to the evacuated post or if family members are authorized to reunite with the employee at the employee's safehaven duty station.

Families earlier evacuated to the U.S. or to an authorized foreign safehaven at USG expense will be permitted to rejoin an employee subsequently evacuated to a duty station in the U.S. However, from an alternate foreign safehaven, travel expenses will be reimbursed only on a cost constructive basis calculated from the evacuated post to the U.S. duty station. (S.R. 631a (1)).

Return to post travel expenses are authorized only after the USG officially permits employees and eligible family members to return. Evacuees should use the same Evacuation Travel Orders issued when they departed post to make the reservations to return to post. Family members not returning at this time may be placed on voluntary SMA. (SR.264.2 (2) exception, 5-15-94).

When employees and/or eligible family members are away from a post on official travel (home leave and return orders, R&R, family visitation travel, emergency visitation travel, temporary duty) at the time an evacuation is ordered, travel expenses may be paid by USG to the safehaven location from the employee/family member's location once the purpose of the original travel orders is completed. In some cases, during an Authorized Departure, the Chief of Mission is granted the authority to permit employees/family members, who have been caught out in this way, to return to post.

When employees and/or eligible family members are away from a post on personal travel when an evacuation is ordered, travel to the safehaven will be on a cost-constructive basis, not to exceed cost of travel from the evacuated post to the safehaven location.

SUBSISTENCE EXPENSE ALLOWANCE (SEA)

SEA is payable the day following arrival at the authorized safehaven point for up to 180 days following the date of the evacuation order. Regardless of the timing of an evacuee's departure from post or the status of the evacuation order, SEA may not be paid beyond the 180th day after the order is issued. SEA payments and benefits are designed only to help offset the direct added expense incurred incident to an evacuation. If family members are not allowed to return to post after expiration of the 180 day period, the employee may request Separate Maintenance Allowance (involuntary) on their behalf. In some cases the employee will be eligible to receive Transitional Separate Maintenance Allowance (TSMA) on behalf of his/her family. In either case the allowance is not automatic. The employee must file an application with the Regional Bureau. (Please note that SEA and Medical Expense Allowance cannot be claimed at the same time. See Evacuee Medical Information in this document for additional information.)

When the U.S. is designated as the official safehaven, the SEA is based on safehaven location per diem. The first evacuee of a family unit is reimbursed for lodging expenses, based on either a commercial or non-commercial rate.

- **Commercial rate:** The first evacuee is reimbursed for 100% of lodging expenses for duration of the evacuation. Meals and incidental expenses are covered at 100% of the local rate for the first 30 days. The rate is reduced to 80% for days 31 through day 180. Commercial receipts for lodging (e.g., hotels, motels, commercially leased houses or apartments, or other transient-type commercial quarters) are required when claiming the commercial rate. (See chart under Calculating Subsistence Expense Allowance.)

- **Non-commercial rate:** For the first 30 days, the first evacuee is paid 10% of the lodging expenses and 100% of meals and incidental expenses. Non-commercial lodging expenses are not reimbursable for days 31-180. However, from days 31-180, 80% of meals and incidental expenses will be paid. No receipts or accounting is required. This rate applies in the absence of commercial receipts (e.g., lodging in government quarters or staying with friends or relatives). (See chart under Calculating Subsistence Expense Allowance.)

When a foreign area is designated as the official safehaven, the SEA is based on the applicable per diem rate of the officially designated safehaven under the same payment guidelines as for the United States.

When an alternate foreign safehaven is approved, SEA can be authorized no earlier than the managing Bureau in the Department receives the request for an alternate safehaven. SEA is based on the lowest of the following:

- The standard CONUS rate;
- The per diem rate for the official safehaven (foreign or U.S.); or
- The per diem rate for the alternate approved safehaven.

SEA will continue to be paid during periods of annual leave and sick leave (except when in medical evacuation status) to employees and family members who are in evacuation status.

SEA will not be paid to employees and eligible family members who are in home leave or R&R status, until such time as they were officially due to return to post. Employees on LWOP qualify for SEA only if evacuated as an eligible family member.

AIR FREIGHT

An air freight allowance of 250 pounds for the first evacuee (200 pounds for the second evacuee, 150 pounds for the third evacuee, and 100 pounds each for the fourth or more family members) will be authorized for both evacuation from and return to post -- in accordance with the current agency travel regulations; DOS Standardized Regulations (DSSR) 631 a (3). In lieu of airfreight, a replacement allowance is authorized if air freight is not shipped from post. The allowance for the first evacuee without family is $250. For the first evacuee with one family member it is $450 and for the first evacuee with two **or more** family members the allowance is $600. Air freight is authorized upon return to post, even if no air freight was shipped from post.

TRANSPORTATION ALLOWANCE

A daily transportation allowance of $25/day per family unit is authorized (receipts are not required).

LEASE COVERAGE

If an evacuation is terminated **and an evacuee elects to return to post (and has not been converted to PCS orders),** a waiver on an advance or reimbursement of expense should be authorized for the unexpired period of the lease up to 30 days or waiver of repayment of an advance.

> *Note: PCS (Permanent Change of Station) Orders take precedence over Evacuation Orders. PCS Orders are enacted at the time the employee initiates travel and can affect evacuation allowances. For additional information please contact the Office of Allowances.*

ADVANCE PAYMENTS

An employee may receive an advance SEA payment to help the employee defray the immediate incidental expenses of his/her evacuation (Authorized or Ordered Departure) and/or the evacuation (Authorized or Ordered Departure) of dependents. The amount of the advance payment is the monetary amount covering a period of up to 30 days as determined appropriate by the authorizing officer. The advance payment may be paid to the employee or a designated representative

HOUSEHOLD EFFECTS (HHE)/PERSONALLY OWNED VEHICLE (POV)

Access to and delivery of stored household effects for evacuees is at personal expense, not Government expense unless evacuees are not allowed to return to post after the 180 days and are placed on involuntary SMA. In the latter case, agency travel regulations governing SMA apply. Shipment of POV is not authorized at USG expense until the employee has a permanent change of station (PSC).

EDUCATION ALLOWANCE

Please see the Family Liaison Office's Internet site for education information useful for parents of school-age children:
http://www.state.gov/m/dghr/flo/rsrcs/pubs/4958.htm

Official safehavens are usually established in the United States, and education allowances are therefore normally not paid. U.S. public schools are available to all residents in the U.S. However, if a child was attending school on an away-from-post education allowance at the time of the evacuation order, the allowance may continue for the remainder of the school year. Likewise, a child who qualified for the Special Needs Education Allowance will be allowed to continue to receive the allowance through the duration of the evacuation or until a legal Individualized Education Plan (IEP) is developed at their public school.

Only in rare cases will the official safehaven be designated in a foreign area. Education allowances will be paid. However, if an individual chooses to safehaven in an alternate foreign safehaven (other than the one designated as official), education allowances will *not* be paid.

Education travel eligibility rules continue to be applied as provided for in Chapter 280, Standardized Regulations, except that the official safehaven displaces the post as the travel destination from school. No SEA benefits are payable for children at their school locations when utilizing either the away-from-post education allowance or educational travel.

For more information and resources on returning to the Washington, DC area with children, please visit FLO's Web site at www.state.gov/m/dghr/flo/reentry/.

TANDEM COUPLES

Tandem couples will each receive evacuation benefits not to exceed an employee's eligibility, but without duplication of benefits for eligible family members. (Each employee is entitled to all other allowances, including transfer allowances and temporary lodging.)

LENGTH OF THE EVACUATION

The initial cable ordering the evacuation declares the evacuation for a specified period of time (normally 30 days). At the end of that period, the Department, in conjunction with the post, reviews the evacuation status to determine whether it should be continued, whether employees should be reassigned, or whether to terminate the evacuation. If the evacuation is not terminated, the status must be reviewed every 30 days. No evacuation will last beyond 180 days since the legislative authority to pay evacuation benefits expires by law. If family members are not allowed to return to post and no reassignment decision has been reached, the post will become an unaccompanied one. At such time, eligible family members may apply for Involuntary Separate Maintenance Allowance (ISMA) or Transitional Separate Maintenance (TSMA) depending on their situation.

More on Subsistence Expense Allowance

The **Department of State Standardized Regulations** apply to all government civilians in foreign areas but each agency may have further implementing regulations. Be sure to check your agency's regulations before applying the DSSR (for Foreign Affairs agencies consult the FAM)

PER DIEM AND MEALS AND INCIDENTAL EXPENSES

SEA is based on the safehaven location per diem rate rather than the continental US (CONUS) per diem rate.

- Only the first evacuee will be reimbursed for lodging expenses based on either a commercial rate (receipts required) or a non-commercial rate (no receipts required).

- All evacuees will be reimbursed a flat amount to cover daily meals and other expenses (M&IE) based on a percentage of the per diem rate.

- The lodging portion of the SEA remains at 100% throughout an evacuation. Evacuees age 18 and over receive 100% of the M&IE the first 30 days. On the 31st day up to 180 days, the M&IE is reduced to 80%. Evacuees under age 18 receive 50% of the M&IE for the first 30 days. On the 31st day up to 180 days the M&IE is reduced to 40%.

- Larger families may request a waiver asking that commercial lodging per diem be increased to 150%. Justification must be provided.

LEASE COVERAGE

When an evacuation has ended and the first evacuee is bound to a lease agreement for lodging at the safehaven, the evacuee is authorized either: (1) expenses for the unexpired period of the lease for up to 30 days or (2) waiver of repayment of an advance. This benefit is only available to evacuees who subsequently return to post.

DAILY TRANSPORTATION ALLOWANCE

A daily transportation allowance of $25 per day per family is allotted to offset costs due to lack of private transportation at safehaven.

AIRFREIGHT REPLACEMENT ALLOWANCE

- Airfreight is authorized when an evacuee leaves post.

- An airfreight replacement allowance is authorized when airfreight is not shipped from post when evacuated. The amounts are $250 for first evacuee without other family members; $450 for first evacuee with one additional family member; and $600 for first evacuee with two or more additional family members.

- Upon return to post, an evacuee is authorized an air freight shipment back to post after the evacuation ends, regardless of whether they came out with a shipment or took the air freight replacement allowance.

Subsistence Expense Allowance Application

For Agencies other than the Department of State: The **Department of State Standardized Regulations (DSSR)** apply to all government civilians in foreign areas but each agency may have further implementing regulations. *Be sure to check your agency's regulations before applying the DSSR (for Foreign Affairs agencies consult the FAM).*

ATTENTION DOS PERSONNEL: SEA payments will be made when the online DS-4095 form http://arpsdir.a.state.gov/eform/formsearch.html is completed and all necessary ORIGINAL documents are submitted and processed.

To expedite processing, forms may be faxed to **(843) 202-3803**. However, the hardcopy original must be delivered to:

> Department of State, Global Financial Operations,
> Charleston Financial Services Center, FM/GFS/F/AO,
> P.O. Box 15008, Charleston, S.C. 29415-5008,
> Attn: Sherry Howard, CAA

Necessary documents: airline ticket, travel orders, hotel receipt or signed lease agreement, taxi receipts, receipt for excess luggage fee, and copy of travel advance.

Receipts are required for all expenses e.g. airline tickets, hotel bills, taxi fare. ORIGINAL documents must be submitted with forms.

Calculation of Meals & Incidental Expenses (M&IE): If travel is more than 12 hours, but less that 24 hours then: First day = 3/4 day of the location traveling to; Lodging plus M&IE for stopover locations en route to safe haven, if applicable; Last day = 3/4 day of location traveling from (when departing to return to Post). All days in between = SEA.

Please Note: The employee is responsible for repaying any SEA balance if reassigned, if evacuation ends abruptly, if issued PCS/TDY/MED/R&R/Home Leave orders, or if lodging changes from commercial to non-commercial.

Should you have questions, please contact Sherry Howard at (843) 746-0722, or via e-mail at HowardSA@state.gov

Calculating the Subsistence Expense Allowance

For official U.S. safehaven calculations (based on U.S. safehaven per diem rates)

Sample tables for Subsistence Expense Allowance (SEA) & Separate Maintenance Allowance (SMA) rates:
By Family Size - Daily Rates   By Family Size - Monthly Rates

Per Diem Rate CONUS   Per Diem Rate Non-CONUS   Per Diem Rate Foreign

L = Lodging portion of per diem.
M&IE = Meals and Incidental Expense portion of per diem.

Table to Apply for Commercial Lodging		
	Days 1 – 30	**Days 31 - 180**
First Evacuee	100% L + 100% M&IE	100% L + 80% M&IE
Each additional person 18 & over	100% M&IE	80% M&IE
Each additional person under 18	50% M&IE	40% M&IE
Table to Apply for Non-Commercial Lodging		
First Evacuee	10% L + 100% M&IE	80% M&IE + no lodging per diem
Each additional person 18 & over	100% M&IE	80% M&IE
Each additional person under 18	50% M&IE	40% M&IE

Checklist for Evacuees Returning to Post

When an evacuation is lifted before 180 days, evacuated employees are expected to return to post in an expeditious manner. Except for those individuals who are converted to PCS orders, there is a grace period of three days (this includes Saturday and Sunday) for the continuation of SEA to allow evacuees to make transportation and travel arrangements for return to post.

An additional seven days may be available for justifiable reasons and in certain circumstances, but the grace period can not be extended past 10 days after the official date of the lifting of the evacuation. The additional seven days are NOT automatic; they must be justified, and the justification must accompany the final travel voucher. To assist in planning a return to post, please double-check the following items:

- Contact your official travel office for assistance with routing, reservations and tickets.

- **Department of State and USAID** employees and family members are required to utilize Carlson Wagonlit. Carlson Wagonlit can do the following:
  - Send your tickets to you by overnight mail;
  - Have the tickets delivered to the airport – this *requires AT LEAST 72 hours of advance notice;* or
  - You can collect your tickets at the Carlson Wagonlit office at Main State (HST), Room 1243.

- The evacuee must supply Carlson Wagonlit with their travel orders containing their fiscal data. NOTE: To avoid penalty in "city pair" or contract fare routes, it is imperative to use the official travel office.
  - **DOS** employees and family members telephone: Reservations (866) 654-5593, after hours and emergencies, (866) 654-5612.
  - **USAID** employees and family members telephone (866) 343-5009.
  - **DEA** families should make travel arrangements through Omega Travel by faxing them a request along with an approved DOJ-501 at (703) 359-3912.

- Arrange for pick-up and shipment of UAB (airfreight). Evacuation Travel Authorizations are used for return travel and shipments.

- o **DOS** families contact the DOS Transportation Office at (202) 647-4140 or toll-free (800) 424-2947.

- o **USAID** families contact the USAID Transportation Office at (202) 712-4848.

- o **DEA** families contact the DEA Family Liaison Specialists at (202) 307-8222 or (202) 307-4241.

- File change of address with local post office.

- **Department of State** employees complete check-out with the Employee Service Center, tel. (202) 647-3432, fax (202) 647-1429, email: EmployeeServicesCenter@state.gov.

- Inform bank, credit cards, etc. of change of address.

- Check validity of Diplomatic Passport and country (your post) visa.

- Employees, check the expiration date of the I.D. card and/or building pass for your agency and request renewal if necessary.

- **DEA** employees turn in headquarters building or parking passes to Security Programs.

- Notify post of arrival plans. Please contact the Post Management Officer (PMO) within the Regional Bureau at the Department of State. The PMO will cable post with the information. See page 6 for telephone list of DOS regional bureaus or contact FLO if you need assistance.

- **USAID** families contact your Desk Officer.

- Upon return to post, file a travel voucher with your post's Budget and Finance office (B&F) to be forwarded to your agency's budget office to reconcile your account.

- Family members who choose not to go back to post immediately may opt for Separate Maintenance Allowance (SMA) http://www.state.gov/m/a/als/faq/5321.htm#sma until returning to post, extending the amount of time they are away from post. This use of SMA does not count as the one-time SMA option per tour. Families in commercial lodging who choose to remain behind to finish the final semester of the school year may apply for the Transitional Separate Maintenance Allowance (TSMA) http://www.state.gov/m/a/als/faq/5321.htm#tsma.

- If you filed for your evacuation allowances using the DS-4095 form, please be sure to close out your account with Charleston (CAA). Sherry Howard, CAA Accountant, will need your final paperwork to both close out your account and to take care of any Subsistence Expense Allowance benefits due to you. It is important that you send her your ORIGINAL return airline tickets, final lodging receipts and other applicable documents (e.g. the paperwork from DOS's Transportation Office showing when your UAB shipment was collected to be sent back to post); be sure to let Sherry's office know the details and dates for everything (if they are not clear on the documents you enclose in your mailing to her).

If you experience difficulty in making arrangements to return to post, please contact the Crisis Management and Support Officer in the Family Liaison Office at (202) 647-1076 or via email at floaskevacuations@state.gov, your PMO, or the support staff for your agency.

Frequently Asked Questions on Evacuation

From the Department of State Office of Allowances *http://www.state.gov/m/a/als/faq/5321.htm*

GENERAL

1. Q: What is the difference between an authorized departure and an ordered departure?

A: Authorized departure merely allows the Chief of Mission greater flexibility in determining which employees or groups of employees may depart, and avoids any negative connotation that might be attached to the use of the term "evacuation." Since the law uses the terms synonymously, there is no difference in benefits now in application of the regulations.

 Note: Once the Under Secretary of State for Management ("M") approves the evacuation status for post—either authorized or ordered—the 180-day clock "begins ticking" (by law, an evacuation cannot last longer than 180 days).

2. Q: Do all US Government agencies subscribe, follow or adhere to the DSSR on evacuations?

A: In accordance with DSSR 645 all civilian agencies have agreed to implement the regulations. In order to ensure fair and consistent treatment of all evacuees, any agency that proposes to deviate from DSSR Chapter 600 must seek Secretary of State approval.

Uniformed military personnel and their dependents are covered separately under the Department of Defense Joint Federal Travel Regulations.

3. Q: How are Members of Household (MOHs) of U.S. direct hire employees affected during an evacuation?

A: MOHs who are U.S. citizens are provided the same evacuation assistance as private American citizens. The "no double standard" policy requires private American citizens be given the same evacuation opportunities/assistance as official Americans. It is the Department of State policy to make available to private Americans any evacuation option planned for the official USG community, when appropriate and feasible. MOHs who are not U.S. citizens are also rendered evacuation assistance, either to the U.S. if documented for entry or otherwise eligible to enter as determined by the post, with assistance from INS as necessary, or to a third country if documented or otherwise determined to be eligible for entry.

4. Q: Who is responsible for travel costs of Members of Household (MOHs)?

A: MOHs are personally responsible for their travel costs. As a general rule, the U.S. Government does not provide funds for evacuating persons other than U.S. Government employees and eligible family members.

5. Q: Where can I find more information on evacuation guidelines for U.S. citizens and other persons for whom the U.S. Government may have a responsibility?

A: The Overseas Citizens Services, Bureau of Consular Affairs provides assistance for U.S. citizens living abroad. You can contact them on the internet at http://www.travel.state.gov

6. Q: Do Members of Household (MOHs) have to leave during an evacuation?

A: No. The Chief of Mission (COM) cannot order the departure of MOHs or other private U.S. citizens.

7. What happens if a Member of Household (MOH) is occupying USG housing - would s/he be required to vacate the house?

A: If the Chief of Mission (COM) feels that the presence of MOHs in post housing could impact post security resources or otherwise affect post operations, then the COM can direct the employee to move MOHs out of that housing.

SAFEHAVEN

8. Q: How does an eligible family member (EFM) select an official safehaven and what is the subsistence expense allowance (SEA) based on?

A: The Under Secretary for Management ("M") consistently designates the United States (the 50 United States and the District of Columbia) as the official safehaven location. Although there is a provision in the DSSR to designate a foreign or non-foreign location, outside the U.S., as the official safehaven, this is rarely if ever done (DSSR 610I). An EFM evacuee should select an official U.S. safehaven, (any of the 50 United States or the District of Columbia) based on where he/she anticipates spending the longest time—such as where children may go to school or where family/ friends reside. Although an evacuee is not required to remain at the official U.S. safehaven, any SEA payments will be based on the per diem rate of the official safehaven location. An official safehaven location (in the U.S) for EFMs may be changed once during an evacuation and SEA payments will then be based on the new official U.S. safehaven. However, any change in safehaven location is at the evacuee's personal expense, except when previously evacuated EFMs are allowed to rejoin their evacuated sponsor (the employee) in Washington, DC (or wherever his/her work assignment location is during evacuation). Be sure to notify the appropriate evacuee locator and accounting offices of any change in address.

9. Q: May an evacuated employee choose to accompany family members to their safehaven point prior to the employee reporting for work at the official safehaven (Washington, D.C. or other U.S. locations as determined by the employee's agency)?

A: Either the post or appropriate bureau or agency head determines whether an employee must report directly to the official safehaven and if any annual leave is authorized. However, the employee's travel at government expense will be cost constructive not to exceed the cost of travel from the post to Washington, D.C. (or other U.S. locations as instructed by the employee's agency, per DSSR 631a.(1)). The employee is not eligible for SEA until he/she arrives at the official safehaven.

10. Q: May an evacuated employee accompany EFMS to their official safehaven if they are unable to travel alone due to special needs or minor age?

A: In accordance with agency travel regulations, an employee or other adult escort (e.g. friend, nanny, other evacuee) may accompany EFMs unable to travel alone to their official U.S. safehaven and receive reimbursement for travel expenses. When an employee elects to have an escort accompany EFMs to their official U.S. safehaven, the escort's travel will be on a cost constructive basis calculated from the evacuated post to the U.S. duty station.

11. Q: What benefits are available for those going to an alternate foreign safehaven?

A: Benefits are only available if the alternate foreign safehaven is approved as in the best interest of the U.S. Government by the Under Secretary of State for Management ("M") following endorsement by the head of agency or designee on behalf of the agency employee's eligible family members (EFMs). The only benefits are cost constructive travel (not to exceed the cost of travel between the evacuated post and the U.S. duty state—DSSR 631a(1)—and limited SEA (based on the lowest of the official safehaven, alternate approved safehaven or standard CONUS per diem rate—currently $99). No education allowance of any type can be authorized, nor are any diplomatic immunities, privileges, or services available at an alternate approved safehaven.

12. Q: May previously evacuated eligible family members (EFMs) join the employee at his or her official safehaven?

A: Yes, EFMs evacuated to a U.S. or authorized foreign safehaven may be permitted to rejoin the employee subsequently evacuated to a U.S. duty station. Transportation costs would be at U.S. Government expense for the family members from their official U.S. safehaven to the employee's U.S. duty station. EFM travel from an alternate approved safehaven to the employee's official U.S. duty station will be on a cost constructive basis not to exceed the cost of travel from the evacuated post to the U.S. duty station (DSSR 631a(1)).

13. Q: May an employee and eligible family member (EFM) children take an authorized departure from post while the employee's spouse remains at post for professional reasons?

A: Yes, an employee and EFM children may be granted an authorized departure from post upon approval of the Chief of Mission, per DSSR 610j. If the employee's spouse and the employee were a tandem couple, the employee's spouse would not be required to evacuate. If the employee's spouse is a locally hired employee at the Mission or is otherwise employed outside the Mission the spouse would not be required to evacuate, for example, if the employee's spouse was employed as a teacher in an international school.

14. Q: May an eligible family member (EFM) on educational travel or "away from post" education allowance, travel to the safehaven following evacuation of a post?

A: Yes. Either the official or alternate foreign safehaven location displaces the foreign post of assignment for travel purposes under education allowance and educational travel as provided in DSSR 633.2 and DSSR 633.4, respectively. Note: Payment of subsistence expense allowance (SEA) is not allowed for children on "away from post" education allowance (DSSR 633.2). Payment of SEA is allowed for children under educational travel only when they are at the safehaven and only for children unmarried and under 21 (see definition of child in DSSR 040 m. (2)

15. Q: Do newly assigned employees/eligible family members (EFMs) not yet arrived at the post qualify for evacuation benefits under DSSR Chapter 600?

A: Under the evacuation benefits law, only employees and dependents who are temporarily away from their foreign post of assignment at the time of the evacuation order are eligible for evacuation benefits if prohibited from returning. However, under the transfer allowance authority, DSSR 245 does allow equivalent benefits to certain newly assigned personnel who are prohibited from proceeding to post. These equivalent benefits provide for SEA payments as well as justified grace periods needed to return to post at the end of an evacuation, and the lease penalty payment.

To qualify for these equivalent benefits the following three criteria must be met on the date of the ordered/authorized departure:
(1) the employee's transfer orders had been issued;
(2) on the date of the ordered/authorized departure order the employee must have been within 60 days of scheduled departure to the new post; and
(3) either (a) the employee's HHE had already been packed out and the employee's residence had been vacated; or (b) the employee transferring from a post in the U.S. has an irrevocable contractual agreement for lease or sale of his or her residence; or (c) the employee has transferred from a foreign post with direct transfer orders (i.e. no home leave, or equivalent, prior to reporting to the new foreign post) and is required by post to vacate residence quarters.
If all three criteria are not met, EFMs are eligible for only Involuntary Separate Maintenance Allowance under DSSR 262.1.

SUBSISTENCE EXPENSE ALLOWANCE (SEA) AND ADVANCE PAYMENTS

16. Q: When do Subsistence Expense Allowance (SEA) benefits commence for evacuees?

A: (1) U.S. Safehaven: SEA benefits will commence from the day following arrival at the official safehaven location, per DSSR 632. No SEA will be paid for travel enroute to official safehaven location.

(2) Alternate Approved Safehaven (outside the U.S.): If an alternate foreign safehaven is approved prior to the EFMs evacuation, SEA benefits will commence from the day following arrival at the approved alternate safehaven location. If an alternate safehaven is approved after evacuees have arrived at that location, SEA will commence no earlier than the date the Department received the request for the alternate safehaven. If the request for an alternate safehaven is denied, no SEA is authorized until the evacuee arrives at the authorized safehaven.

17. Q: What is the maximum time period an employee may receive SEA payments?

A: Evacuation status is authorized by the Under Secretary of State for Management in 30-day increments, up to a maximum of 180 days, per DSSR 623f.

18. Q: What happens to an employee's allowances and differentials during the period of SEA payments?

A: Post differential and post allowances terminate as of the close of business on the day an employee commences travel under orders for emergency evacuation, per DSSR 621.2(a) and (f). "School at post" education allowance terminates without financial penalty, per DSSR 621.2(d)(1). "School away from post" education allowance may continue until the end of the school year, per DSSR 621.2(d)(2). Employees should check their earnings and leave statements for differential and allowance payments that should have been terminated. The employee is responsible for refunding any overpayments. (Refer to DSSR 532e. for termination of post differential in those cases when the employee is away from post on detail or leave and his or her departure from post is ordered.)

19. Q: Do SEA payments stop while an evacuated employee is on temporary duty (TDY)? What happens to the employee's evacuated eligible family members (EFMs)?

A: To meet the needs of the service, employees in evacuation status may be assigned TDY to another location. However, during the TDY period, when the employee is receiving TDY per diem, the payment of SEA for the individual is suspended, per DSSR 635(a). SEA may be resumed when the TDY ends. If there are EFMs of this employee in evacuation status, one family member receiving SEA becomes the first evacuee and thus receives lodging per diem. The family is not disadvantaged since the total SEA benefit package is reduced by only one MI&E allowance when the employee departs on TDY status.

20. Q: When an employee is evacuated after eligible family members (EFMs) have been evacuated at an earlier date, is the employee entitled to SEA under DSSR 632.1, at the full amount for the first evacuee or at the amount for an additional family member?

A: When the employee is evacuated later than EFMs, the employee may be treated as the first evacuee or simply as an additional family member. The DSSR allows for eligible family member(s) and the employee to be at different safehaven locations, however, there can be only one "first evacuee" under the formula (whether commercial or non-commercial) at any given point in time. Note: For reimbursement for larger quarters, only family members residing with the first evacuee at his/her safehaven are counted for this special consideration.

21. Q: May employees and EFMs on home leave or on R&R receive SEA payments?

A: Employees and EFMs cannot receive SEA while on home leave or in R&R status per DSSR 632.4(a). If away from post at the time of the evacuation order, the employee either must return to post or declare his/her intention to do so before any family member can qualify for evacuation benefits. Transportation may be authorized to the official safehaven location. SEA may not commence for evacuees until each arrives at the authorized safehaven and providing the employee has commenced official travel to the duty station (either to safehaven or return to post). Determination of the exact date may also in some circumstances depend on the date the employee or EFMs members were officially due to return to post.

22. Q: If Permanent Change of Station (PCS) travel orders have been issued prior to an employee/EFMs, which orders take precedence?

A: PCS travel orders always take precedence over any other type of travel orders, including evacuation orders. An evacuee's travel should therefore be charged to his/her PCS orders. An evacuee may, however, be eligible to receive SEA benefits if the evacuation occurs prior to his/her originally scheduled PCS travel. When EFMs depart post under evacuation orders, and the employee subsequently departs post under PCS orders, all evacuation benefits will cease for EFMs when the employee's PCS travel begins.

23. Q: What if EFMs have been evacuated and the employee later joins the evacuated family on a different type of travel order such as R&R or home leave?

A: Employees cannot receive SEA. However, SEA continues for family members previously evacuated per DSSR 632.4.

24. Q: How is it determined whether the commercial or non-commercial rate for SEA applies, and can an employee draw SEA at the commercial rate and EFMs members draw SEA at the non-commercial rate?

A: Commercial or non-commercial applies to the type of lodging the "first evacuee" occupies. Per DSSR 632.1, "There can only be one 'first evacuee' at any given time, except as per DSSR 632.4(b) ('Tandem Couples')". Only the first evacuee (this may be the employee or eligible family member) is reimbursed for up to 100 percent (or up to 150 percent for special family compositions) of the lodging portion of the per diem rate for his/her safehaven location. All other eligible family members get only a percentage of the meal and incidental expense portion of the per diem rate of the first evacuee's safehaven location. If the first evacuee submits a commercial lodging receipt, then the commercial rate formula applies. If a commercial lodging receipt is not submitted, then the non-commercial rate formula applies. Both formulas are shown in a separate document entitled, " Evacuation Payments Worksheet (EPW)", in DSSR 960. Note: If evacuees stay in commercial lodging that does not include furniture and/or cost of utilities, these additional costs may be reimbursed as part of lodging (receipts must be submitted with lodging receipts).

25. Q: If the family composition requires more than one hotel room or larger quarters is there flexibility to allow reimbursement above the commercial rate maximum of 100% of the lodging portion of the safehaven?

A: The first evacuee may be reimbursed up to 50% above the lodging maximum when using the commercial rate, per DSSR 632.1(b). Special consideration is given to the following family compositions: (1) First evacuee plus one (non-spouse family member, age 18 and older); (2) First evacuee plus one (non-spouse family member of opposite gender, age 12 and over); (3) First evacuee plus two (one non-spouse family member, age 18 and older; or one non-spouse family member, opposite gender, age 12 and older); (4) First evacuee plus three (one non-spouse family member, age 12 and over); and (5) First evacuee plus four or more family members.

26. Q: What if I have a special family composition not included as one of the five listed in the previous question?

A: Requests for other special family considerations should be submitted through your agency to the Director, Office of Allowances (A/OPR/ALS), U.S. Department of State, Washington, D.C. 20522-0103, per DSSR 632.1(b).

27. Q: Upon termination of an evacuation order, is there a grace period for continuation of SEA until the day an evacuee returns to post? What if the employee is being transferred and not returning to post?

A: Upon termination of an evacuation order, an employee or EFM will continue to receive SEA for an automatic grace period of three (3) days except when the full 180 days has expired, per DSSR 635. For the employee not returning to post, only the three (3) days are allowed as long as he/she has not commenced travel under an assignment order to another duty location. For employees/family members returning to the evacuated post, an additional seven (7) days may be authorized due to transportation delays. Evacuees must provide a statement on their travel voucher justifying the additional seven (7) days required to arrange for return travel to post, such as airline reservations and air freight pick up. Other reasons of a personal nature do not qualify the evacuee for SEA for the additional days. Under no circumstance can SEA payments be made to exceed the 180-day "clock".

28. Q: Is my nanny or caregiver eligible for SEA?

A: Unless the nanny or caregiver is an eligible family member (EFM), the answer is "no." However, he/she may be the designated representative (DSSR 610f) who is named by an employee for the purpose of caring for, escorting, or receiving monetary payments on behalf of an EFM.

29. Q: Is there any other provision under the evacuation payments if I need further help with unexpected expenses related to evacuation?

A: Yes. Per DSSR 615, an employee may be paid in advance of the normal payday when, in the judgment of the authorizing officer, payment is required to help defray the immediate expenses incident to an evacuation. The advance payment shall be for a maximum of 30 days based on the rate of compensation including any allowances or post differential the employee was entitled to immediately prior to the evacuation order. The advance payment may be made at any time after the date the evacuation has been ordered but no later than 30 days after this date.

30. Q: What work assignments may an employee expect while on evacuation?

A: Evacuated employees may be assigned to perform any work considered necessary or required to be performed during the evacuation period, per DSSR 625.1. Failure or refusal to perform assigned work may be a basis for terminating further evacuation payments and taking disciplinary action, per DSSR 625.2.

31. Q: An evacuated employee was assigned TDY to another post directly from the evacuated post. When are post differentials terminated and initiated in this situation provided the losing and the gaining posts grant them?

A: Per DSSR 532a(3), the Post Differential terminates for the employee's permanent post the day he/she departs on emergency evacuation orders. The employee will receive the post differential until evacuated from his/her permanent post. The employee will receive the post differential for the TDY post commencing with the 43rd day at post. The employee will not receive SEA payments because the employee has not been evacuated to a safehaven site.

32. Q: An evacuated employee was assigned to the Department and was receiving SEA payments. Subsequently the employee was assigned TDY to a post that has a differential. When will the employee's SEA be terminated and the differential payments initiated?

A: The employee's SEA payments are terminated at the time of departure from Washington, D.C. to the TDY post. The employee will receive the TDY post's differential commencing FROM the 43rd day at this post unless the employee is at a DSSR 920 footnote "N" post for the first 42 consecutive (NOT cumulative) days of detail. If that pertains, then once the 42 days consecutive are obtained, a look back to day one is allowed and all 42 days are paid.

33. Q: When are SEA payments terminated?

A: Entitlement to SEA payments end on the earliest of the following dates, per DSSR 635:
• the date the employee commences travel under an assignment order to another duty station outside the evacuation area;
• the effective date of transfer when the employee is already at the post to which transferred;
• the date of separation;
• the date specified by the head of agency;
• 180 days after the evacuation order is issued; or
• the date the evacuee commences return travel to post.

When a departure (evacuation) order is terminated and evacuees are allowed to return to post, entitlement to SEA payments ends on the day that return to post is authorized. Normally a grace period of three, not to exceed ten days, is granted during which SEA may continue to be paid while an evacuee is making arrangements to return to post. The grace period taken must be justified on the employee's travel voucher (i.e. that the extra days were necessary to arrange return to post). SEA payments are limited to 180 days, the grace period included.

34. Q: What happens after an evacuation has terminated and the post becomes unaccompanied, meaning family members can no longer go to post?

A: Employees whose EFMs have been in temporary commercial lodging should apply for Transitional Separate Maintenance Allowance. Employees whose EFMs have been in non-commercial lodging should apply for Involuntary SMA. In lieu of Involuntary SMA for children in grades K-12, employees may consider the "away from post" education allowance option (see DSS 276.23 for details). Remember that since SMA payments cannot be made retroactively, the employee should make every effort to submit Standard Form (SF-190) before the evacuation ends in order to take advantage of these benefits.

TRAVEL ALLOWANCES

35. Q: Are evacuated employees and/or family members granted unaccompanied air baggage (UAB) for their departure from post?

A: Yes, Department of State employees/EFMS are allowed UAB, per 14 FAM 613.3-1, as follows: First person traveling (employee), 250 lbs.; second person traveling, 200 lbs.; third person traveling 150 lbs.; fourth or more persons traveling 100 lbs. Employees of other agencies should consult their applicable travel regulations for agency-specific weight allowances.

36. Q: What if this authorized UAB shipment cannot be arranged at the time of the employee's and EFMs' evacuation departure?

A: For Department of State employees: in lieu of an airfreight allowance from post, an airfreight replacement allowance may be granted to help defray costs of items normally part of the authorized airfreight shipment that must be purchased at the safehaven location.

The flat rates are:
- first evacuee without family: $250;
- first evacuee with one family member: $450;
- first evacuee with two or more family members: $600.

No receipts are required for this allowance, per DSSR 631a.(3). Note that the maximum airfreight replacement allowance is $600, while the maximum UAB allowance could exceed 600 lbs., depending on family size. Employees of other agencies should consult their applicable travel regulations for airfreight allowances in this situation

37. Q: If an employee receives the above allowance, is the employee entitled to ship UAB back to post upon the termination of evacuation?

A: Yes, the standard UAB shipment weights cited above apply, per DSSR 631a.(3). Employees of other agencies should consult their applicable travel regulations for airfreight allowances in this situation.

38. Q: If an employee ships UAB for an EFM to the U.S. safehaven, may UAB be shipped from the U.S. safehaven to the employee's next duty station where the EFM rejoins the employee?

A: Yes. However, the UAB is shipped using the employee's PCS orders, not the evacuation orders. Total weight of UAB shipped from the safehaven to the next post and from the employee's current duty station to the next post may not exceed the UAB weight authorized on the PCS orders. Employees of other agencies should consult their applicable travel regulations for airfreight allowances in this situation.

39. Q: A spouse of an employee is considered as emergency personnel and works in an EFM position at an Embassy that is on ordered departure. When the evacuation terminates and the post becomes unaccompanied, the EFM position is to be abolished and the spouse must then leave post for safehaven in the U.S. Under what kind of orders may the spouse travel and what allowances may be granted?

A: The spouse travels under evacuation orders to the involuntary SMA location, per DSSR 264.1. The employee may receive involuntary SMA commencing the day after the spouse arrives in the U.S. The spouse is not entitled to transitional separate maintenance allowance (TSMA) because the spouse did not leave post during the evacuation and did not receive SEA payments.

40. Q: After an evacuation terminates and the post becomes unaccompanied, what benefits are available to EFMs? For example, an employee's EFMs want to move from their safehaven in Washington, DC to an alternate involuntary SMA location in the U.S. What allowances may be granted the employee for the family members?

A: For the Department of State, Human Resources (HR) will fund the EFMs' travel from the safehaven (in this case Washington, DC) to the involuntary SMA location and will issue SMA orders that designate Washington, D.C. as the alternate point of origin to the SMA location, per DSSR 264.1. The EFMs may ship their evacuation UAB to this location and may access their HHE at government expense. Employees of other agencies should check with appropriate officials within your agency for guidance.

41. Q: After a period of time, if an unaccompanied post is declared accompanied and the employee's EFMs want to return to post, what allowances may be granted the employee?

A: For the Department of State, Human Resources (HR) will fund the EFMs' travel back to post, provided the return to post is prior to the employee's final 90 days before PCS, per DSSR 264.2(2). Employees of other agencies should check with appropriate officials within your agency for guidance.

42. Q: After an evacuation terminates and EFMs may return to post, an employee requests that EFMs remain at the safehaven site for two months before returning to post. What allowances may be granted an employee whose EFMs remain at the safehaven on voluntary SMA under the exception at DSSR 264.2(2)?

A: Following termination of an authorize/ordered departure an employee may elect voluntary SMA at the official safehaven for eligible family members previously eligible for SEA payments under DSSR Chapter 600 and for whom round-trip travel expenses have already been authorized. The employee may be permitted to then terminate this voluntary SMA and these eligible family members may be permitted to return to post provided return travel to post does not occur during the employee's last 90 days at a post of assignment. No additional expenses for travel, access to goods in storage, shipment of household effects or other such SMA-related expenditures may be incurred on their behalf. Note this election of voluntary SMA does not count as the "one change of election during a tour of duty" as normal voluntary SMA would.

43. Q: An employee is evacuated from post leaving a pet behind to be cared for by servants or friends. While at safehaven in Washington, DC and assigned to the Department the employee's tour was curtailed. The employee went on home leave and then proceeded to a newly assigned overseas post on PCS. The employee wants to send the pet to the new post. May this shipment be done at government expense?

A: Shipping a pet is the employee's responsibility so this expense is not covered by transfer orders. However, the cost of shipping a pet may be claimed under the miscellaneous expense portion of the Foreign Transfer Allowance. The maximum amount of reimbursement for all allowable expenses without receipts is $1,000 for an employee with family. Greater amounts up to two weeks of the employee's salary at the time of entrance on duty at the new post may be granted with receipts or other acceptable evidence justifying the amount claimed for expenses. The current maximum is two weeks of the employee's salary or that of a GS-13, step 10, whichever is less. Costs in excess of these maximums are considered personal expenses that may be claimed, in some cases, as moving expenses for income tax purposes.

EDUCATION ALLOWANCES

44. Q: Are any other special education benefits available to evacuees and their families?

A: Usually education allowances are not granted for children evacuated from post to a safehaven in the U.S. if accompanied by a parent. However, if prior to evacuation, a child was attending school in the U.S and was receiving the "away from post" education allowance, the rate authorized for the evacuated post may continue for the remainder of the school year, per DSSR 633.2. There is no entitlement for SEA payments for children on "away from post" education allowance.

45. Q: Does the "away from post" education allowance continue during the evacuation period?

A: The education allowance continues until the end of the current school year and the safehaven location replaces the foreign post of assignment as the destination for travel within the education allowance. No payment of SEA is authorized for any period of time covered by the "away from post" education allowance (DSSR 633.2).

46. Q: May an EFM on educational travel and "away from post" education allowance, travel to the safehaven following evacuation of a post?

A: Yes. The safehaven location replaces the foreign post of assignment for travel purposes under education allowance and educational travel as established in DSSR 633.2 and DSSR 633.4. Note: Payment of SEA is not allowed for children on "away from post" education allowance (DSSR 633.2). Payment of SEA is allowed for children under educational travel only when they are at the safehaven and only for children unmarried and less than 21 years old. ("Child" is defined in DSSR 040m. (2).)

47. Q: May an EFM student travel at government expense from the student's safehaven site (the student's school is located in the U.S. to the home leave address of the employee to join the family on home leave?

A: Yes, this travel is authorized, per DSSR 284. The EFM student would have been on the home leave orders of the employee had the student been able to return to post, a return that was precluded owing to the evacuation.

48. Q: During an evacuation, EFM students at "away from post" schools (boarding schools) continue to receive the "away from post" education allowance until the end of the school year, but do not receive SEA during the spring school break. When the school term ends, what options are available for students?

A: After the school years ends, the students may travel to a safehaven location using the remaining travel monies from the "away from post" education allowance. They may then begin receiving evacuation payments (SEA). The reason is that if the evacuation continues into the succeeding school year, students would not be eligible for an education allowance when the new school year begins. However, the employee may pay any required deposit or tuition payment for the first semester of this school year from personal funds with the understanding the employee risks not receiving reimbursement for this expenditure.

Should the evacuation terminate during the school year and EFMs are permitted to return to post, the Office of Allowances within the U.S. Department of State will update the "at post" and "away from post" education allowances which could subsequently be used to fund educational expenses for that school year (assuming the EFMs met the requirements of DSSR 270). Posts may not pay a deposit or down payment for tuition payment for the succeeding school year while the post is in authorized or ordered departure status.

If after termination of the evacuation, the post is declared "unaccompanied", and there is no parent residing in the United States then the student is permitted to use the "away from post" education Allowance to attend boarding school in the U.S., per DSSR 276.23 and DSSR 276.3.

49. Q: What options are available to boarding school students on "away from post" education allowance who would normally be permitted to travel back to post during an evacuation? May they travel to the safehaven at government expense? What are their options should the evacuation terminate and they wish to return to post for the summer break?

A: If there is no parent already evacuated to the U.S., the employee may designate a relative's or other adult's residence, or the city in which the boarding school is located, as the safehaven. The employee has the option of paying for travel from the boarding school to the safehaven location from personal funds and thereby "save" the remaining travel monies of the "away from post" education allowance for later use in the event the evacuation terminates and EFMs are permitted to return to post. Please refer to the question/answer 5 above regarding the options available should the evacuation continue into the succeeding school year.

50. Q: What options are available to college students who would normally be eligible for educational travel during an evacuation? The safehaven normally replaces the post for computing educational travel, but is it permissible to delay using educational travel to post in anticipation of the imminent termination of evacuation and thus retain the option of the EFM student returning to post during the summer break?

A: If there is no parent evacuated to the U.S., the employee may designate a relative's or other adult's address or the city in which the college or university is located as the safehaven, per DSSR 610 I. The student, if under the age of 21, may begin receiving SEA payments after the student arrives at the safehaven. Parents have the option of paying for travel from the educational institution to the safehaven location themselves, thereby retaining the educational travel trip for later in the summer in the event the evacuation terminates and the post becomes accompanied again. If a student is already living in an off-campus apartment or similar housing at the end of the school year, then this accommodation will not be considered commercial lodging for purposes of SEA payments.

51. Q: An employee is assigned to an unaccompanied post. The employee's EFM student, age 18, is enrolled in college but cannot visit the employee using educational travel. In these circumstances, may the employee receive involuntary SMA for the EFM student?

A: Yes, the employee may receive involuntary SMA in lieu of educational travel in this instance, until the EFM student becomes 21 years of age.

52. Q: Some families have EFM students in boarding school and/or college in different locations that in addition terminate the school term on different dates. Must all family members proceed to the same safehaven site or may EFMs of the same family declare different and separate safehaven sites? If they are permitted to travel to different safehaven sites, what is the basis for the MIE payments?

A: EFMs are not required to travel to the same safehaven site. They may travel to alternate safehaven sites either in or outside of the US, per DSSR 631a.(1). However, there can be only one authorized first evacuee, so the employee must designate one EFM as that evacuee, per DSSR 632.1. Travel of all other EFMs to alternate safehaven sites is calculated on a cost-construct basis using the safehaven of the first evacuee as the basis for this cost calculation. Lodging, if applicable and MIE payments are based on the rates established for the first evacuee.

53. Q: At the termination of an evacuation and a declaration that EFMs may return to post, an employee's EFMs nonetheless wish to remain in the U.S. so that the EFMs students may complete the final term of the current school year. What allowances may be granted the employee?

A: If the EFMs are residing in commercial quarters, the employee may receive a maximum of 90 days of Transitional SMA (TSMA) to allow the EFM students to complete the final semester of the school year, per DSSR 262.3b. If the EFMs are residing in non-commercial quarters, the employee may be granted Voluntary SMA (VSMA). However, EFMs may not return to post if, at the end of the school year, the employee is within 90 days of PCS.

TRANSITIONAL SEPARATE MAINTENANCE ALLOWANCES (TSMA)

54. Q: An employee returns to the on evacuation orders to join his previously evacuated family two weeks before the end of the 180-day evacuation period. At the end of the 180-day evacuation period, the employee's post is declared "unaccompanied." At this time the employee takes two weeks of annual leave and then returns to post. May the employee receive TSMA on behalf of this family for the two weeks following the termination of the evacuation, even though the family was not separated?

A: Yes, TSMA may be granted for this period because the employee was maintaining quarters at the post during this temporary absence from post, per DSSR 265.3. TSMA in this case would be for the purpose of "transitioning " the family from commercial quarters occupied during the evacuation to permanent quarters because the post has been declared "unaccompanied", per DSSR 262.3a.

55. Q: An employee's EFMs were evacuated six months ago and have been living in a furnished apartment. The evacuation has been terminated and the post is declared as "unaccompanied". The EFMs would like to move to a less expensive rental house. Is this move permissible while still retaining TSMA eligibility?

A: Yes, this move is permissible provided that the less expensive commercially leased rental house is a "transition" residence prior to occupying permanent quarters. However, if this less expensive rental house is intended to be a permanent residence then the employee may not receive TSMA, but may receive involuntary SMA, per DSSR 262.3a.

56. Q: After an employee's EFMs move into a permanent residence are they still eligible to receive TSMA until they receive their full HHE shipment?

A: TSMA payments terminate the earliest of the dates that are set forth in DSSR 266.4, i.e. (a) the date the employee commences travel under transfer orders from the evacuated post or the date of transfer if no travel is to occur under the transfer orders; (b) the final day of the authorized period of the TSMA; (c) the date the complete HHE shipment is received by the employee's EFMs; (d) the date the EFMs occupy non-commercial quarters; (e) the date EFMs occupy permanent quarters.

If the employee's EFMs move into a permanent residence before they receive their full HHE shipment then their TSMA will terminate before they receive that shipment. Alternately, if they receive their full HHE shipment before occupying permanent quarters TSMA will terminate on the date they receive that shipment.

57. Q: An employee ships some HHE from post to EFMs on TSMA in the U.S. If the HHE does not arrive within 60 days may TSMA be extended beyond this period?

A: Usually TSMA under DSSR 262.3a may be paid for a maximum of 60 days, but an additional 30 days may be allowed, with agency approval based on extreme or unusual circumstances, per DSSR 267.1b.(2). The employee at post should submit an SF-1190 (Rev. 12/2006) to the appropriate agency official before the end of the 60-day period to request an extension of TSMA payments for the additional 30 days. An example of "extreme or unusual circumstances" would be a situation in which the employee made reasonable efforts to ship the HHE to his EFMs in the U.S. but the shipment did not arrive in this time period.

58. Q: What are the TSMA rates?

A: TSMA rates for days 1–30 are $100/day for 1-2 family members and $120/day for 3 or more family members; for days 31-60 $70/day for 1-2 family members and $80/day for 3 or more family members; for days 61-90 $50/day for 1- 2 family members and $60/day for 3 or more family members.

59. Q: Is TSMA taxable?

A: No, TSMA is not subject to federal or state income taxes, per DSSR 054.1.

60. Q: How are TSMA payments initiated, received, and then terminated?

A: An employee submits a SF-1190 (Rev. 12/2006) that is processed and approved, per DSSR 264.3. Payments are made to the employee by payroll and continue until the employee submits a SF-1190 requesting termination of the allowance. The employee should submit this termination notice upon the initial occurrence of any of the following events, per DSSR 266.4: (a) the date the employee commences travel under transfer orders from the evacuated post or the date of transfer if no travel is to occur under the transfer orders; (b) the final day of the authorized period of the TSMA; (c) the date the complete HHE shipment is received by the employee's EFMs; (d) the date the EFMs occupy non-commercial quarters; (e) the date the EFMs occupy permanent quarters.

61. Q: Once TSMA is terminated what options remain for the employee?

A: The employee may submit an SF-1190 to the appropriate agency requesting involuntary SMA for each EFM and specifying in box #18 of the SF-1190 the reason for the request; namely, that the post is unaccompanied and travel to post has been denied to the employee's EFMs, per DSSR 264.1. An employee's minor EFMs are covered by involuntary SMA until the age of 21 (age 18 for voluntary SMA with the exception of those over 18 who are in secondary school). Involuntary SMA is paid according to the following rates: $5,800 for one child; $8,700 for 2 or more children; $9,900 for one adult; $13,400 for one adult and one additional family member; $15,200 for one adult and two or three family members; and $17,700 for one adult and four or more additional family members, per DSSR 267.1a.

62. Q: What other allowances are available once TSMA terminates and the post is declared unaccompanied?

A: If the employee has EFM students in grades K through 12 then the employee may request either involuntary SMA or the applicable "away from post" education allowance. The applicable "away from post" education allowance is that specified for the employee's post of assignment, per DSSR 276.23. The "away from post" education allowance option may be chosen unless child is going to school in the U.S. and the parent (natural, adoptive, or step) resides in the U.S. also (rare exceptions noted at DSSR 276.3). If this is the case, the employee is not eligible for the "away from post" education allowance. The logic is that the child could live with the parent and attend public school free of charge. This same prohibition does not apply if the child is going to school in a foreign country and the parent resides in the same foreign country because school wouldn't be free in the foreign country as in the U.S. The only restriction on the "away from post" education allowance when a child is going to school in a foreign country is at DSSR 277.2. Room and board could not be paid to the parent if the child lived with that parent outside the U.S. (rare exceptions noted at DSSR 277.2).

63. Q: May an employee receive TSMA for EFMs who have been evacuated to an alternate approved foreign safehaven?

A: EFMs are eligible to receive TSMA, per DSSR 264.3, if they are occupying temporary commercial quarters, per DSSR 264.3. They are not eligible for TSMA if they are residing in non-commercial quarters. If they are not eligible for TSMA they may be eligible for either involuntary SMA or "away from post" education allowance (see the previous Q&As on this topic).

Note: EFMs are officially considered to be residing in the U.S. on involuntary SMA. However, the employee may request that the alternate approved foreign safehaven of the EFMs be designated as the official involuntary SMA site. If this request is approved by the appropriate agency official an education allowance may be paid for eligible EFM students unless they have traveled to a school under the educational travel authority within the previous 12 months (DSSR

262.5). (Department of State employees should make this request to the Executive Director of the appropriate regional bureau.)

64. Q: What constitutes "non-commercial" quarters?

A: "Non-commercial" quarters are those that are not commercially leased or rented. Employees and EFMs living with family, friends, etc. would be considered living in "non-commercial" quarters, per DSSR 632.1.

65. Q: An employee is to be reassigned within a month after the termination of evacuation orders. May his EFMs living in the U.S. receive TSMA for this period?

A: The employee's EFMs are eligible for TSMA while they reside in temporary commercial quarters. However, if the employee is expecting to be transferred to the U.S. within a month the subsistence expense portion of the Home Service Transfer Allowance (HSTA) may provide a better benefit for the employee than the TSMA. If TSMA is approved, EFMs will not be eligible for the HSTA unless "official transportation was authorized permitting those family members to join the employee at the new post of assignment in the U.S. , per DSSR 252.8.

SEPARATE MAINTENANCE ALLOWANCES (SMA)

66. Q: At what age must voluntary SMA terminate for a dependent child?

A: Voluntary SMA must be terminated on a child's 18th birthday, unless the child is attending secondary school or is determined to be incapable of self-support (due to physical or mental impairment), per DSSR 264.2(1)c.

67. Q: If an employee's spouse is in Washington , D.C. on voluntary SMA and the employee is evacuated from post to the safehaven site of Washington , D.C. is SMA terminated?

A: The SMA is not terminated. When the employee is evacuated, it is considered a temporary absence from post. Per DSSR 265.3 the grant shall continue during the absence of the employee from the post provided the employee maintains quarters at the post, unless terminated under the provisions of DSSR 266.2 or 266.3 (Transfer or Separation).

68. Q: At what age must involuntary SMA terminate for a dependent child?

A: Involuntary SMA must be terminated on a child's 21st birthday, unless the child is determined to be incapable of self-support (due to physical or mental impairment), per DSSR 267.1a. A child who is in post-secondary school/college and not currently working is not considered to be incapable of self-support.

69. Q: If an eligible family member on SMA travels to the post at personal expense will the Department assume any responsibility in the case of an emergency involving the EFM?

A: When an EFM travels to the employee's post at personal expense, the family member is considered a private citizen visiting the country. As such, that family member is not eligible for any allowances or benefits paid on behalf of family members of USG civilian employees, including any medical emergency evacuation travel, per DSSR 261.2. However, if the employee has not previously used the one change of option per tour (voluntary SMA), the employee might be permitted to do so once an emergency arises.

70. Q: When an employee receiving SMA is transferred to another post does the SMA automatically continue?

A: When an employee is transferred, SMA must be terminated, per DSSR 266.2. The employee must then elect to apply for SMA at the new post or have family members included on the travel orders (if an accompanied post).

71. Q: Following the termination of evacuation, may an employee's EFMs remain at their safehaven on voluntary SMA and then return to post later?

A: Following termination of an authorized or ordered departure an employee may be elect voluntary SMA at the official safehaven for the EFMs previously eligible for SEA payments and for whom round-trip travel expenses have already been authorized, per DSSR 262.2. However, an EFM who has been living in an official overseas safehaven and wishes to remain there must reaffirm that this location meets Department of State security standards. The employee may be permitted to terminate this voluntary SMA and EFMs may be permitted to return to post provided return travel to post does not occur during the employee's final 90 days at post, per DSSR 264.2(2). This SMA is not to be considered the "one change of option" during a tour of duty.

72. Q: Who may officially authorize approval or disapproval of an employee's SMA request?

Employees must check with their agencies for officials who are authorized to approve/disapprove SMA. Within the foreign affairs agencies, the following officials are authorized to approve SMA requests for their respective agencies (see 3 FAM 3232 and 3FAH-1 H3232.2 for additional instructions):
State Department--Executive Director of the appropriate bureau

U.S. Agency for International Development (USAID) --See ADS 477
USDA--FAS -- Director, International Services Staff
Department of Commerce--Director, Office of Foreign Service Human Resources
Only the following officials can disapprove SMA applications for their respective foreign affairs agencies:
State Department--Deputy Assistant Secretary for Human Resources
USAID--See ADS 477
USDA--FAS -- Deputy Administrator, Foreign Agricultural Affairs
Department of Commerce--the Director General or Appropriate Secretarial Officer.

73. Q: What procedure must an employee follow in submitting an SMA request?

A: The employee must complete and sign the SF-1190 (Rev. 12/2006) that contains a statement certifying the veracity of the request. Employees from agencies other than State should contact their respective headquarters for guidance. If an employee is requesting SMA on behalf of a spouse, the spouse must also sign the SF-1190 so that the spouse is aware of the request for SMA on his/her behalf.

74. Q: Is an employee's EFM student eligible to receive educational travel while on SMA?

A: No. When EFM students are on voluntary SMA they are not eligible for other allowances, including educational travel.

75. Q: Is an employee's EFM student eligible to receive an education allowance while on involuntary SMA?

A: If a foreign area is designated as the official SMA (involuntary) location, an EFM student may be eligible for an education allowance, per DSSR 262.5. Unless specifically designated otherwise by the head of agency, EFMs on SMA (voluntary) are considered to be officially residing in the U.S. and can attend U.S. public schools free of charge. However, if SMA is granted for the convenience of the government (involuntary) and a foreign area is designated as the official SMA location, the EFM student authorized to reside at that location may be authorized an education allowance within the applicable "away from post" education allowance established in the regulations at post unless the child has traveled under the Educational Travel authority within the previous 12 months.

76. Q: The spouse of an employee at a foreign post is residing in the U.S. on SMA. When the employee leaves (transfers from) post the SMA will be terminated. How much Home Service Transfer Allowance is the employee

eligible to receive? Further, should the subsistence portion of this allowance include the spouse as well?

A: Family members for whom an SMA was authorized while the employee was posted in a foreign area are not considered members of the family for computing the Home Service Transfer Allowance unless official transportation was authorized permitting those family members to join the employee at the new post of assignment in the U.S., per DSSR 252.8. Therefore the employee is granted a Home Service Transfer Allowance as a single employee, not as one with EFMs.

SINGLE PARENTS / TANDEM COUPLES

77. Q: If a single parent employee is assigned to a position designated by the post as an "emergency" position will the employee's children be evacuated when ordered departure for EFMs is declared? If so, what allowances will the EFM children receive?

A: Yes, the EFM children will be evacuated. They will be eligible for the customary evacuation benefits, per DSSR 630. If there is a single child, this child would be considered as the "first" evacuee of the family unit and would receive SEA based on the safehaven's per diem rate, including the actual lodging charge up to the lodging limit. If there are other children evacuated they would receive a percentage of the safehaven's locality M&IE rate according to the formulas set forth in DSSR 632.1.

78. Q: When tandem couples depart post on evacuation orders, how are their SEA payments determined?

A: Tandem couple employees will each receive evacuation benefits not to exceed an employee's eligibility, but without duplication of benefits for family members on their orders. Both employees are considered to be the first evacuee, per DSSR 632.1. (This is based on tandem couple eligibility for all other allowances, including transfer allowances and temporary lodging, on a per person employee basis. However, only 50% of the lodging allowance is granted if the tandem couple is sharing lodging.)

79. Q: In the case of a tandem couple with EFM children and only one parent being evacuated, on whose orders should the evacuated children be placed?

A: EFM children of tandem couples should be placed on the evacuating employee/parent's orders, per DSSR 610j.

80. Q: How does a tandem couple evacuated to the same official safehaven submit their receipts under the commercial rate formula for lodging?

A: Provided the evacuated couple is residing in the same commercial quarters, they should submit their vouchers together. Reimbursement would be granted as "first evacuee" to each for lodging (one-half of the commercial lodging charge for each if sharing quarters) as well as M&IE, per DSSR 632.1(b).

SHIPMENTS AND PROPERTY CLAIMS

81. Q: May POVs be shipped from the post to the safehaven point at government expense?

A: No, POV shipments are not authorized, per 631b. However, a safehaven transportation allowance of $25 per day is authorized. No receipts are required.

82. Q: May an employee have access to HHE while on evacuation status?

A: Access to, delivery and return to storage of household effects for evacuees is at personal expense, not Government expense, per DSSR 631b.

83. Q: If an employee does not ship UAB from post during an evacuation and subsequently receives the airfreight replacement allowance, may the employee ship UAB back to post after the evacuation?

A: Yes. The airfreight replacement allowance is only in place of the UAB from post, per DSSR 631a.(3).

84. Q: What if an employee is assigned to a new post, but all personal effects remain at evacuated post?

A: If the household effects (HHE), unaccompanied air baggage (UAB), or privately owned vehicle (POV) are packed out, they will be shipped as soon as it is possible to do so. If the employee is not present at post, the post's Management section will be responsible for packing and shipping the employee's effects.

85. Q: What should an employee do if personal property has been lost or damaged at post?

A: The employee may file a claim with the government for loss or damage of property, but not until such time that the post verifies the loss. The claim is filed through the employee's agency. Foreign Affairs Agency (State, USAID, FAS, FCS, APHIS) personnel must complete Form DS 1620, "Claims for Loss of or

Damage to Private Personal Property." The Management Officer at post files an investigative report. A cable from post verifying the loss or damage may be acceptable. If possible, the employee should present travel orders, inventory, receipts, and bills of lading. If not possible, the employee should file the claim with whatever information is available.

86. Q: What is the maximum amount that may be claimed for lost and damaged personal property?

A: There is a maximum amount for certain categories of items claimed, with a total maximum amount of $100,000 for damages sustained during an evacuation. It is essential to carry personal loss insurance and to keep this insurance current for even if the employee's private insurance company refuses to pay (owing, for example, to acts of war) and the USG becomes the primary insurer, the Department's claims office will rely on the insurance valuation when assessing the employee's claim.

87. Q: Who should the employee contact to begin this claims procedure?

A: Department of State employees should contact the Transportation and Travel Management Division Claims Section (A/LM/OPS/TTM/CL), room 1245, HST, as follows:

Joan Padilla, team leader, tel. # 202-736-7061 PadillaJoanD@state.gov
Carolyn Baker, tel. # 202-736-7154 BakerCP@state.gov
Jewel Woodard, tel. #202-736-7156 WoodardJD@state.gov

TERMINATION OF EVACUATION

88. Q: What is the period of validity for evacuation orders?

A: Under normal circumstances, evacuation orders are valid for up to nine months from date of issuance. However, if an employee still at post is reassigned to another post, evacuated family members are not allowed return travel within 30 days of reassignment.

89. Q: How are SEA payments and travel vouchers administered at the termination of an evacuation period?

A: For SEA payments: For Department of State employees, Resource Management (RM) will provide a miscellaneous obligation document (MOD), which summarizes evacuation payments. The evacuee must certify this document "as is," or with modifications and returned to: RM/GFS/DFS/FO/A, SA-

15, Room 5000, Washington, DC 22209) with appropriate documentation of changes within 10 days.

Evacuees should provide justification for any time between the termination of the evacuation and initiation of return travel to post. Justification includes making necessary travel arrangements. Documentation is essential so that RM can complete the final audit of SEA payments.

For travel payments: For Department of State, evacuees returning to post should file their travel vouchers with the post budget officer. The post computes the travel vouchers based on current per-diem rates and forwards them to RM for certification of payment. Evacuees who are not returning to post should file their vouchers with their regional bureau. These vouchers will be computed by the bureau and forwarded for certification of payment.

90. Q: What happens if an evacuation ends and an employee is liable for paying a lease penalty?

A: If an employee or designee signs a lease at the safehaven and is subsequently ordered to return to post, or if an evacuation terminates and the post subsequently becomes unaccompanied, then the employee's agency may waive the refund due the Government on an advance or reimbursement of lodging expenses incurred not to exceed 30 days. This lease coverage may not extend beyond the 180-day evacuation payment limit. See DSSR 632.4c. INVOLUNTARY SEPARATE MAINTENANCE ALLOWANCE (ISMA)

91. Q: If families are separated for the convenience of the government, why are all of our housing expenses not covered like they would be at post?

A: The purpose of both Voluntary and Involuntary Separate Maintenance Allowances is to help defray the additional expenses associated with maintaining family members elsewhere than at post. The allowances are not meant to fully cover a family's expenses. The rates for Involuntary SMA are based on data provided by the Bureau of Labor Statistics for the average cost of maintaining a household in the continental United States (CONUS). The costs include average rent, utilities, miscellaneous furnishings and supplies. Because individual circumstances vary and most people do not live in the "average" area in CONUS, the rates may not fully cover each family's costs.

92. Q: If the post allows only spouses (or spouses and children under 5), but a spouse cannot go because there are children (or older children), is the employee then eligible for ISMA?

A: In addition to fully unaccompanied status, the Department of State has added the category of "partially unaccompanied" posts, i.e. Only adult EFMs, or adult EFMs and small children, are permitted. If minor children (under the age of 18 years) may not proceed to post and are therefore eligible for ISMA, a parent or step-parent may remain at the separate household to care for them and would also be eligible for ISMA. Questions on unusual circumstances should be sent to the Director of the Office of Allowances at allowanceso@state.gov.

93. Q: What happens if the family has been on ISMA and the post status changes back to accompanied, but because we are in the middle of the school year it is not advisable for the family to move? Can the employee continue to receive ISMA, or does it shift back to voluntary? If it shifts to voluntary, do I need to fill in another SF-1190?

A: When an unaccompanied post becomes safe enough for EFMs to return and the status changes to at least partially accompanied, the higher ISMA rates will continue for 90 days. Before the 90-day period ends, the employee will need to submit a new SF-1190 (Rev. 12/2006) to continue either ISMA (if partially unaccompanied and the employee's EFMS may not return) or VSMA (if the employee's EFMs could return but choose not to).

94. Q: If a family member fails to get a medical clearance, is the employee eligible for involuntary. Is there a reporting requirement with regard to the medical condition? Do you have to continue to demonstrate the medical need for ISMA? If there is no longer justification for a medical ISMA, would the EFM still be eligible for voluntary SMA?

A: An employee may receive Involuntary SMA for a family member if that family member is prevented from proceeding to post due to a medical condition. The Office of Medical Services or other competent medical authority must certify in writing that the family member's medical condition prevents the family member from going to post. That certification could be a memo, cable or email that is attached to the SF-1190, Foreign Allowances Application, Grant and Report. If the family member's medical condition changes and the family member may proceed to post, but chooses not to, the employee is eligible for Voluntary SMA. The employee must immediately submit the new SF-1190 documenting that the EFM's status has changed. Please note that an employee may only elect VSMA for family members once during a tour of duty and the employee and family member must be separated for at least 90 days. The change of election may not take place during the employee's first or last 90 days during the tour of duty. (DSSR 264.2(2)).

95. Q: I am assigned to an accompanied post and my family members are with me. Unfortunately, one of my EFMs has developed a medical condition

and the Regional Medical Officer has stated that he cannot remain at post with me. He will return to the United States to get medical treatment for the remainder of my tour. May I put him on SMA? What type of SMA would he receive - voluntary or involuntary?

A: You should submit an SF-1190 to your post management officer to apply for SMA on behalf of your EFM. Please note on the SF-1190 that you are applying for involuntary SMA because the Office of Medical Services has limited your EFM's medical clearance in such a way that he is not permitted to reside with you at post. Please attach documentation from the RMO or Office of Medical Services indicating that needed medical facilities are not available at post and therefore the EFM's medical clearance is being limited. You would then be eligible for ISMA on behalf of your EFM.

96. Q: What support is available to families on involuntary separation?

A: The Family Liaison Office (FLO) within the Department of State takes the lead on providing support to families on separate maintenance, both voluntary and involuntary, as well as those on evacuation status. FLO is setting up a network of support for families on involuntary separate maintenance and wants you to contact FLO with your email address and telephone number. FLO has developed a resource book and a website for unaccompanied tours on the internet website http://www.state.gov/m/dghr/flo/c14521.htm.

97. Q: How does ISMA relate to travel of separated families?

A: Travel for children of separated families described in 3FAM 3753 does apply to children on involuntary separate maintenance allowance. Since children are not allowed to visit an unaccompanied post, travel would have to be to an alternate location.

EVACUATION

98. Q: My family and I were evacuated six months ago and I have been leasing a furnished apartment in a high rise building. I'd like to move to a rental house that is less expensive. Can I do this and still receive Transitional Separate Maintenance Allowance (TSMA)?

A: Yes, you and your family can move to a less expensive commercially-leased rental house as long as it is a "transition" residence prior to occupying your permanent residence. If this move to the less expensive rental house is intended to be your permanent residence, then you will not be eligible for TSMA but only eligible for the "regular" Involuntary SMA rates in the Department of State Standardized Regulations (DSSR) section 267.1.

99. Q: Even after my family moves, they will have a lot of extraordinary expenses (like furniture rental) until they receive the full Household Effects (HHE) shipment. When will my TSMA payments stop?

A: As stated above, if this residence is considered temporary ("transitional") and commercially-leased, then TSMA payments may be paid for up to 60 calendar days awaiting your full HHE shipment. TSMA is intended to help defray the extraordinary expenses your family will experience during this transition period between the end of evacuation and the beginning of Involuntary SMA. The daily TSMA rates for days 1 through 30 are $100 per day for 1 or 2 eligible family members; and $120 per day for 3 or more eligible family members. The daily TSMA rates for days 31 through 60 are $70 and $80, respectively. These are not per person rates but per family rates.

100. Q: Even if my husband, who is still at post, ships some of our HHE right away; it will still probably take more than 60 days to get here. How do I extend the TSMA to 90 days?

A: TSMA may be paid for a maximum of 60 days with an additional 30 days allowed following agency approval based on extreme or unusual circumstances. The employee at post should submit an SF-1190 to the appropriate agency official ahead of the end of the initial 60 day period to request an extension of TSMA payments for the additional 30 days. One example of extreme or unusual circumstances is that the employee has made every effort to get the full HHE shipment to the family but the full shipment has not been delivered due to restrictions or difficulties beyond the employee's control.

101. Q: Are the TSMA amounts different for days 61 through 90?

A: Yes, the TSMA rates for days 61 through 90 are $50 for 1 or 2 eligible family members and $60 for 3 or more eligible family members.

102. Q: Is TSMA taxable?

A: No. TSMA is considered the same as SMA which is not subject to federal or state income taxes.

103. Q: What if I get some of our furniture from storage and also have some of our HHE shipped. Would I still be eligible for TSMA after I get the items from storage, but until the HHE arrives?

A: If you are still in temporary commercial lodging, you are eligible for TSMA for up to 60 days or until your "full" HHE is delivered.

104. Q: How do you know when my "complete" HHE has arrived?

A: The employee should submit an SF-1190 to the appropriate agency official to inform them of the date the family received the "complete" or "full" HHE. TSMA must terminate on the date the full HHE is delivered to the family. Please note that the USG is not responsible to move an HHE shipment from a temporary residence to the subsequent (permanent) residence, therefore, it would be good to coordinate delivery of HHE to a permanent residence. Please note that TSMA terminates when the earliest of several possible events occurs (see DSSR 266.4) and therefore could stop even before the "complete" HHE is delivered if one of those other conditions applies.

105. Q: Is TSMA paid automatically once it commences?

A: Yes. Once an employee has submitted an SF-1190 and it has been processed, payments commenced via the payroll process continue automatically until the employee submits an SF-1190 to the appropriate agency official to terminate the allowance. The employee should submit the form as soon as any of the following occur: date the employee commences travel under transfer orders from the evacuated post or date of transfer when no travel by the employee under the transfer order is involved; date the authorized period for Transitional SMA ends; date the complete Household Effects (HHE) shipment is delivered to family; date the family members occupy non-commercial quarters; date the family members occupy permanent quarters.

106. Q: Once TSMA is terminated, what are my options?

A: You will submit the SF-1190 to the appropriate agency official requesting Involuntary SMA for each family member and cite in box 18 of the SF-1190 the reason for the request [that the post is unaccompanied and transportation to post has been withheld for family members]. Remember, involuntary SMA extends to children until they reach age 21 (age 18 is the limit for voluntary SMA unless the child is in secondary school). Although amounts may change in the future, the current annual amounts for Involuntary SMA are $5,800 for one child; $8,700 for 2 or more children; $9,900 for one adult; $13,400 for one adult and 1 additional family member; $15,200 for one adult and 2 or 3 family members; and $17,700 for 1 adult and 4 or more additional family members.

107. Q: Is there anything else available following termination of TSMA?

For your school age children (grades K through 12), according to DSSR 276.23, the employee may request the applicable "away from post" education allowance [for the employee's post of assignment] for his/her child in lieu of involuntary SMA. The restrictions are that (1) a parent (natural, adoptive, step) cannot reside in the United States if the child will attend school in the United States (this restriction does not apply if the child and parent live in the same foreign country); and (2) the child cannot have traveled under the educational travel authority within the previous 12 months.

108. Q: What if my family is at an alternate approved foreign safehaven at the end of the evacuation. Can I get TSMA for my family members in the foreign area?

A: If your family members are occupying temporary commercial quarters then they are eligible to receive TSMA. However, if they are in non-commercial quarters they are not eligible for TSMA. If they are not eligible for TSMA, they are eligible for either involuntary SMA or away from post education allowance (see Q&A 9 and 10 for details). Note: Although family members are officially considered to reside in the U.S. on Involuntary SMA, if an employee has extenuating family circumstances, he/she may ask the appropriate agency official to "officially" designate a foreign area for Involuntary SMA. If a foreign area is "officially" designated for Involuntary (as opposed to voluntary) SMA, an education allowance (based on the DSSR 920 at post rate for the designated SMA location) may be paid for a child on Involuntary SMA at that location. (For State Department, the agency official would be the Executive Director of the appropriate regional bureau).

109. Q: Can you give me examples of non-commercial quarters?

Non-commercial quarters are considered private residences such as living with family, friends or others in a location which is not commercially leased or rented.

110. Q: I understand you're not supposed to receive regular SMA unless you will be in that status for at least 90 days. I expect my spouse to be reassigned in a month or so. Am I still eligible for TSMA?

Your family is eligible for TSMA as long as they are in temporary commercial lodging. However, if the employee is going to be transferred shortly to the U.S., you may wish to weigh the immediate benefit of TSMA versus the subsistence expense portion of the Home Service Transfer Allowance for family and employee once the employee gets back to the U.S. If TSMA is used, the family will not be eligible for HSTA "unless official transportation was authorized permitting those family members to join the employee at the new post of assignment in the U.S." (DSSR 252.8).

Evacuation Contacts

ALL AGENCIES AT POST

Family Liaison Office
Room 1239, HST (Main State)
Department of State
2201 C. St., NW
Washington, DC 20520
Telephone: (202) 647-1076 or toll-free (800) 440-0397
FAX: (202) 647-1670
Email: flo@state.gov ; Web site: http://www.state.gov/m/dghr/flo

Employee Consultation Services (ECS)
Columbia Plaza, SA-1, H-246.
Telephone: (202) 663-1815
FAX: (202) 663-1456
Email: MEDECS@state.gov

U.S. DEPARTMENT OF STATE

Subsistence Expense Allowance Payments
Charleston Financial Service Center (CFSC)
Sherry Howard
Telephone: (843) 746-0722
Mailing address:
1969 Dyess Avenue
Charleston, SC 29405
Claims:
Room 1244, Main State
Telephone: (202) 736-7648

Employee Services Center
Room 1252, Main State
Telephone: (202) 647-3432
Fax: (202) 647-1429
Email: EmployeeServicesCenter@state.gov

Housing (AAFSW)
Room 1252, Main State
Telephone: (202) 647-3573
Web site: http://www.aafsw.org

Transportation
> Room 1248, Main State
> Telephone: (202) 647-4140
> Or toll-free (800) 424-2947

Carlson Wagonlit Travel
> Room 1252, Main State
> Department of State reservations and help desk Mon-Fri: 7:30am and 5:30pm
> Telephone: (866) 654-5593
> 'After-hours' & emergencies: (800) 383-6723
>
> Room E-1113 at George P. Shultz NFATC (FSI)
> FAX (866) 597-3743
> Web site: http://www.cwgt.com

DOS Regional Bureaus

AF/EX - African Affairs
> Supervisory Post Management Officer Rick Weston
> Telephone: (202) 647-1351

EAP/EX - East Asian and Pacific Affairs
> Supervisory Post Management Officer Geraldine Kam
> Telephone: IVG 8-879-2264
> Email: KamGL@state.gov

EUR/EX - European & Eurasian Affairs
> Supervisory Post Management Officer Michelle LaBonte
> Telephone: (202) 736-1512

NEA/SCA/EX - Near Eastern Affairs
> Supervisory Post Management Officer Georgia DeBell
> Telephone: (202) 647-1519

WHA/EX - Western Hemisphere Affairs
> Supervisory Post Management Officer Officer Kristen Skipper
> Telephone: (202) 647-4473

U.S. AGENCY FOR INTERNATIONAL DEVELOPMENT (USAID)

Please Note: When filing for evacuation allowances, USAID Employees should first check with your bureau. Send completed forms to: USAID/M/FM/CMP, 1300 Pennsylvania Avenue, NW, Washington, DC 20523-7700.

Counseling
Employee Consultation Services (ECS)
Columbia Plaza, SA-1, H-246.
Telephone: (202) 663-1815
FAX: (202) 663-1456
Email: MEDECS@state.gov

USAID Social Worker
Martha Rees
Telephone: (2020) 712-0891
Email: mrees@usaid.gov

USAID Human Resources Bureau Contacts

AFR Africa Bureau
Michelle Walker
Telephone: (202) 712-5579
Email: miwalker@usaid.gov

ANE Asia & Near East Bureau
Cristal King
Telephone: (202) 712-0592
Email: CKing@usaid.gov

EE Europe & Eurasia Bureau
Sanna Solem
Telephone: (202) 712-0656
Email: ssolem@usaid.gov

LAC Latin America & Caribbean Bureau
Angel Mason
Telephone: (202) 712-1868
Email: amason@usaid.gov

USAID Transportation (shipping)
M/AS/TT, Room 2.09, RRB
1300 Pennsylvania Avenue, NW
Washington, DC 20523-4800

Transportation Counselors:
Andrew James (202) 712-0227
Williford Thomas (202 712-1675

Carlson Wagonlit at USAID
Website: http://www.cwgt.com
Telephone (for USAID employees & family members): (202) 866-343-5009

DRUG ENFORCEMENT ADMINISTRATION (DEA)

Foreign Personnel Support Unit/International Ops

Family Liaison Specialists:
Deborah Rice Thomas
Telephone: (202) 307-8222
Fax: (202) 307-3620
Email: Deborah.R.Thomas@usdoj.gov

Patricia Gonzalez
(202) 307-4241
Fax: (202) 307-3620
Email: Patricia.B.Gonzalez@usdoj.gov

FEDERAL BUREAU OF INVESTIGATION (FBI)

Office of International Operations (OIO)
Steve McPeak
(202) 324-9275
Fax: (202) 324-5741
Email: stevenmcpeak@ic.fbi.gov

B. Fiby Gaid
(703) 324-4299
Fax: (202) 324-5741

LaWanda Robinson
(202) 324-5332
Fax: 324-5741

FOREIGN AGRICULTURE SERVICE (FAS)

International Services Staff
(202) 720-2741

FOREIGN COMMERCIAL SERVICE (FCS)

Human Resources
Fay Fivner, telephone: (202) 482-3397
Rose Reedy, telephone: (202) 482-7851

HOMELAND SECURITY

Immigration & Customs Enforcement

Helen Campbell
Telephone: (202) 732-8152
Fax: (202) 305-7793
Email: Helen.campbell@dhs.gov

Vernesia Middleton
Telephone: (202) 732-0399
Email: vernesia.middleton@dhs.gov

UNITED STATES SECRET SERVICE

International Programs

Yolanda Torres
Telephone: (202) 406-5575
Email: yolanda.torres@usss.dhs.gov

Personnel Division

245 Murray Drive, Building 410
Washington, DC 20223
Telephone: (202) 406-5800

Emergency Preparedness

(202) 406 5888Treasury

Human Resources

Joy Charles,
(202) 622-1577

Sherri Coleman,
(202) 622-1043

General information

(202) 622-2000

VOICE OF AMERICA/ INTERNATIONAL BROADCASTING BUREAU (VOA/IBB)

Human Resources

Joanne Lusby or Corrine Appleton
Telephone: (202) 619-311

General Resources

A MESSAGE FROM THE OFFICE OF MEDICAL SERVICES

MEMORANDUM

United States Department of State
Medical Director
Department of State and the Foreign Service
Washington, DC 20520

To: Foreign Service and Foreign Affairs Agencies Returnees
From: Laurence G. Brown, MD Medical Director

Welcome back to the United States. I am grateful for your extraordinary service to the Department in these troubling times.

Your sudden return has probably been hectic, disruptive, and stressful. I want you to know that our staff is ready to assist you as you get re-settled.

The Health Unit at Columbia Plaza (SA-1) room L-206 is available for assessment of acute illnesses and referral to local medical resources. Referrals can also be made to other MED offices as needed, say for medical evacuation or tropical medicine questions. The hours of operation are 8:30 am until 4:00 pm, Monday through Friday.

Mental Health Services works with the Family Liaison Office to arrange briefings and community meetings, and to provide support through informal group discussions on the various community, individual, and family problems associated with evacuations. You will be notified of these meetings.

The Employee Consultation Service (ECS) is a confidential service for Department of State employees and family members and offering consultations, counseling and support groups. The clinical social workers at the ECS have extensive experience in family counseling. The office is located in Room H-246, SA-1, (202) 663-1815.

If urgent assistance is required over a weekend, please call the Emergency Room at a hospital near you for help.

EVACUEE MEDICAL INFORMATION

All employees and family members on Subsistence Expense Allowance (SEA) who seek medical exams should schedule procedures as early as possible during the evacuation period. Once the evacuation order is lifted, it is expected that employees and family members will return to post expeditiously.

Payment for medical coverage provided to eligible US citizen employees and their dependents overseas does not apply to employees and family members while they are in the United States on assignment, home leave, or other travel. The exception is an illness, injury, or medical condition connected with overseas service and the employee or family member is otherwise eligible for treatment. Such a situation is rare - an emergent medical condition for which medevac was approved at the time of ordered departure or a serious condition which occurred overseas necessitating the revocation of a medical clearance, disallowing an individual's return to post. The need to obtain medical care following the SEA period does not in itself warrant the issuance of medical per diem orders.

Employees and family members are encouraged to use the time in the U.S. to obtain medical or dental assistance. The preferred plan providers offered by your Federal Employee Health Benefit Program (FEHBP) reduce your out-of-pocket expenses by choosing facilities and providers who participate in the Plan's preferred provider organization (PPO). You can also reduce up front charges and paperwork if your PPO provider files the insurance claim on your behalf.

Check with your insurance carrier to find out which local facilities and providers participate in your insurance carrier's PPO arrangements. Ask your medical provider if he or she participates in your health plan when you make the medical appointment.

Many insurance programs also provide non-FEHB benefits, for example, dental services, eyeglass examinations and supplies provided at pre-negotiated discounts. Check the brochure or call your carrier.

The FEHBP plans most commonly used by our FS population are listed below. Information is on the Office of Personnel Management's Web site at http://www.opm.gov/insure/health/.

Foreign Service Benefit
Telephone: (202) 833-4910
Fax: (202) 833-4918
Website: http://www.afspa.org

BlueCross BlueShield
Telephone worldwide/collect: (804) 673-1678
Toll-free in the U.S. (800) 848-9766
Website: http://www.fepblue.org

Government Employees Hospital Association(GEHA)
Telephone: Toll-free (800) 821-6136; (816) 257-5500
Fax (816) 257-3233
Website: http://www.geha.com

Mail Handlers Benefit Plan
Tel: (800) 410 7778
Website: http://www.firsthealth.com/smfh/login.do

Please note: *Subsistence Expense Allowance (SEA) and Medical Expense Allowance cannot be claimed at the same time and on the same set of orders.* Each separate set of orders must detail an obligation and a set length of time. MED will not medically travel anyone while in the United States on SEA. Employees and/or EFMs on medical expense orders during an evacuation from post in which the medical order ends prior to the termination of the post evacuation must submit a TA order for SEA entitlement. A copy of the medical order must be forwarded with the SEA application, TA order and any commercial lodging receipt (s) dating from the end date of the medevac . In the event that an SEA evacuee develops a serious medical condition that requires further medical evaluations or treatment, MED/Foreign Programs should be notified for medical clearance to return to post.

HOTELS AND LODGING IN THE WASHINGTON, DC AREA

When looking for lodging at your local per diem rate, an excellent source is the following GSA web site at http://www.gsa.gov. Click on the state where you plan to reside and select the city or county. Click on the "Property List" to find a list of hotels that offer rates within government per diem.

The American Foreign Service Association (ASFA) has a site for Extended-Stay Housing Providers at: http://www.afsa.org/ads/extstay/index.cfm

For hotels in the Dulles Airport area:
http://www.expedia.com/hotels/United_States_of_America-201/Virginia-248/Dulles-184089_01.asp?CCheck=1&

Corporations that assist in finding lodging according to family specific needs and within per diem, at no extra cost to the evacuee

BridgeStreet Corporate Housing Worldwide
Tel: toll-free (800) 278-7338
http://www.bridgestreet.com/

Charles E. Smith
Tel: toll-free (877)902-0832
http://www.smithliving.com

Corporate Apartment Specialists, Inc.
Tel: toll-free (800) 914-2802
http://www.corporateapartments.com

Gilmore Group
Tel: (877) 844-6224
http://www.gilmoregroup.com

Oakwood Apartments
Tel: toll-free (888) 998-3265
http://oakwood.com

HOTEL/SUITE CHAINS IN GREATER METRO DC

A

AmeriSuites
(800) 833-1516
http://www.amerisuites.com

B

Best Western
(800) 780-7234
http://www.bestwestern.com

C

Candlewood Suites
(877) 266-3539
http://www.candlewoodsuites.com

Choice Hotels International[1]
(877) 424-6423
http://www3.choicehotels.com

Country Inn and Suites
(888) 201-1746
http://www.countryinns.com

D

Days Inn
(800) 329-7466
http://www.daysinn.com

E

Extended Stay America
(800) 804-3724
http://www.extendedstay.com

H

Hilton[1]
(800) 445-8667
http://www.hilton.com

Holiday Inn
(800) 465-4329
http://www.ichotelsgroup.com

Hyatt[2]
(800) 233-1234
http://www.hyatt.com

M

Marriott[3]
(800) 932-2198
http://www.marriott.com

Microtel Inns and Suites
(800) 771-7171
http://www.microtelinn.com

My Suite Solutions
(888) 580-0505
http://www.mysuitesolutions.com

S

Sheraton
(888) 625-5144
http://www.starwoodhotels.com/sheraton/

SuiteAmerica
(800) 784-8431
http://www.suiteamerica.com

Summerfield Suites
(877) 999-3223
http://www.wyndham.com

---

[1] *Doubletree, Embassy Suites Hotels, Hampton Inn, Hampton Inn & Suites and Homewood Suites*
[2] *includes Hawthorne Suites*
[3] *Springhill Suites, Renaissance Inn, Residence Inn, Towne Place Suites, Courtyard and Fairfield Inn*

W
Wingate Inn
(800) 228-1000
http://www.wingateinns.com

HOTELS FREQUENTLY USED BY FS PERSONNEL

One Washington Circle Hotel
http://www.thecirclehotel.com
(800) 424-9671

The River Inn
http://www.theriverinn.com
(888) 874-0100

State Plaza Hotel
http://www.stateplaza.com/
(800) 424-2859

CAR RENTAL INFORMATION

Alamo - www.alamo.com (800) 462-5266
Avis - www.avis.com (800) 230-4898
Budget - www.budget.com (800) 527-0700
Dollar - www.dollar.com 800-800-3665
Enterprise - www.enterprise.com (800) 736-8222
Hertz - www.hertz.com (800) 654-3131
National - www.nationalcar.com (800) 227-7368
Flexcar - www.flewcar.gom (877) 353-9227
Zipcar - www.zipcar.com (866) 494-7227

SCHOOL DISTRICT CONTACT INFORMATION - WASHINGTON, DC AREA

District of Columbia	www.k12.dc.us
Information:	Tel: (202) 442-4044
Special Education:	Tel: 202) 442-5468

Northern Virginia	

City of Alexandria www.acps.k12.va.us
Information: Tel: (703) 824-6600
Gifted Program: Tel: (703) 824-6680
Special Education: Tel: (703) 824-6650

Arlington County www.arlington.k12.va.us
Community Services &
Public Information: Tel: (703) 228-6005
Gifted Program: Tel: (703) 228-6160;
Special Education: Tel: (703) 228-6046

Fairfax County www.fcps.k12.va.us
Office of Community Relations: Tel: (703) 246-2991
Office of Student Services &
Special Education: Tel: (703) 246-7780
Gifted/Talented Section: Tel: (703) 876-5272

City of Falls Church www.fccps.k12.va.us
Public Information: Tel: (703) 248-5600
Gifted and Talented Programs: Tel: (703) 248-5603
Special Education Coordinator: Tel: (703) 248-5630

Loudoun County www.loudoun.k12.va.us
Public Information Office: Tel: (703) 791-8720

Prince William County www.pwcs.edu
Community Relations Office: Tel: (703) 791-8720

Stafford County www.pen.k12.va.us/Div/Stafford/
Information Tel: (540) 658-6000

Maryland

Anne Arundel County   www.aacps.org
Information: Tel: (410) 222-5000
Special Schools: Tel: (410) 222-5410/5411
Gifted and Talented   Tel: (410) 222-5430

Howard County   www.howard.k12.md.us
Public Information Office: Tel: (410) 313-6600
Email address for general
questions or comments: publicinfo@hcpss.org
Special Schools Web site
(special needs/magnet schools): hwww.howard.k12.md.us/

Montgomery County   www.mcps.k12.md.us
Information: Tel: (301) 279-3391
Special Education: Tel: (301) 279-3135
Gifted and Talented &
Magnet Programs: Tel: (301) 279-3163

Prince George's County www.pgcps.pg.k12.md.us
Communications: Tel: (301) 952-6001
Special Education: Tel: (301) 952-6336
Talented and Gifted: Tel: (301) 322-1729

OTHER SCHOOL RESOURCES ON THE INTERNET

Washington Area Schools Information
www.state.gov/m/dghr/flo/rsrcs/pubs/1987.htm

International Baccalaureate (IB) Programs
www.state.gov/m/dghr/flo/rsrcs/pubs/1992.htm

Returning to Washinton (childcare, education, special needs, independent schools, pre-school registration and more).
www.state.gov/m/dghr/flo/reentry/